T0223003

QUANTUM NON-LOCALITY AND RELATIVITY

For Vishnya, who always believed in it

Quantum Non-Locality and Relativity

Metaphysical Intimations
of Modern Physics

Third Edition

Tim Maudlin

WILEY-BLACKWELL

A John Wiley & Sons, Ltd., Publication

This edition first published 2011
© 2011 Tim Maudlin

Blackwell Publishing was acquired by John Wiley & Sons in February 2007. Blackwell's publishing program has been merged with Wiley's global Scientific, Technical, and Medical business to form Wiley-Blackwell.

Registered Office
John Wiley & Sons Ltd, The Atrium, Southern Gate, Chichester, West Sussex, PO19 8SQ, United Kingdom

Editorial Offices
350 Main Street, Malden, MA 02148-5020, USA
9600 Garsington Road, Oxford, OX4 2DQ, UK
The Atrium, Southern Gate, Chichester, West Sussex, PO19 8SQ, UK

For details of our global editorial offices, for customer services, and for information about how to apply for permission to reuse the copyright material in this book please see our website at www.wiley.com/wiley-blackwell.

The right of Tim Maudlin to be identified as the author of this work has been asserted in accordance with the UK Copyright, Designs and Patents Act 1988.

Library of Congress Cataloging-in-Publication Data
Maudlin, Tim.
 Quantum non-locality and relativity : metaphysical intimations of modern physics / Tim Maudlin. – 3rd ed.
  p. cm.
 Includes bibliographical references and index.
 ISBN 978-1-4443-3126-4 (hardback) – ISBN 978-1-4443-3127-1 (paperback)
 1 Physics–Philosophy. 2. Quantum theory. 3. Relativity (Physics) I. title.
 QC6.M3612 2011
 530.01–dc22
                                        2011001784

A catalogue record for this book is available from the British Library.

This book is published in the following electronic formats: ePDFs 9781444396959; Wiley Online Library 9781444396973; ePub 9781444396966

Set in 10/12pt Sabon by SPi Publisher Services, Pondicherry, India

1 2011

Contents

Preface to First Edition

I state my case, even though I know it is only part of the truth, and I would state it just the same if I knew it was false, because certain errors are stations on the road to the truth. I am doing all that is possible on a definite job at hand.

Robert Musil

If the introductory chapter of a book is the overture to the ensuing score, a brief, undeveloped melange of themes and leitmotifs destined to appear again and again, the preface serves as program notes. Here may one find some small account of the events which propelled the project; some acknowledgment of the many friends who encouraged and nourished it; and some explanation of idiosyncratic elements which arise from the author's own peculiarities. And as an intriguing introduction may encourage the reader to warm to the subject, so the successful preface may inspire some sympathy and understanding in the reader for the author's plight, for the many compromises, lapses, and errors that attend the writing of a book. So how did this book come about?

In October 1990, John Stewart Bell succumbed quite suddenly and unexpectedly to a hemorrhage of the brain. Anyone who had studied Bell's works mourned the passing of an incisive intellect; those who had had the pleasure of discussing the knotty problems of quantum theory with him felt even more sharply the loss of a figure of inspiring integrity, clarity, and humor. Here at Rutgers, Renée Weber suggested that we honor Dr Bell's memory with a symposium on his work. David Mermin treated us to a non-technical exposition of Bell's Theorem, Shelly Goldstein spoke of the relationship between Bell's work and that of David Bohm, and Professor Weber recounted some parts of her recent interview with Bell. My part was to be a short discussion of the compatibility between Relativity theory and the violation of Bell's inequality.

When I originally agreed to the assignment, I thought that I knew just what I was going to say: Relativity has been interpreted in two quite different ways, as forbidding superluminal effects and as demanding Lorentz invariance, and one must sort out how to construe Relativity before one can address the question of compatibility with quantum theory. But after a few days I realized that another construal of Relativity was available (no superluminal signals), then another (no superluminal energy transmission), then yet another (no superluminal information transmission). Since all of these interpretations of Relativity were provably non-equivalent, this situation posed a straightforward analytical task: how do the various interpretations relate to one another and how does each fare if Bell's inequality is violated? This manuscript is my attempt to work through that analytical problem.

In writing the book I have been constantly surprised by the variety and beauty of the interconnections between these various questions. But I have been even more impressed by Bell's deep and steady understanding of the problematic. Over and over I found some terse passage in Bell's work to contain exactly what needed to be said on a subject, the decisive pronouncement. I have often felt that whatever is of value in this book could be found in Bell's "The Theory of Local Beables" (1987, ch. 7), and have consoled myself that this book will have served a great purpose if it does no more than encourage people to read Bell with the care and attention he deserves.

My foremost goal in composing the book has been to make it comprehensible to the non-specialist. The sparks which fly when quantum theory collides with Relativity ignite conceptual brushfires of particular interest to philosophers, problems about causation, time, and holism, among others. Unfortunately, much of the work done by philosophers presupposes a considerable amount of familiarity with the physics. This is particularly sad since the physics is not, in most cases, very complicated. I fear that many readers may be frightened off from the topic by unnecessary formalization, so I have tried to keep the mathematical complexity of the discussion to a minimum. But on the other hand, I have not wished to drop to the level of vague metaphor which sometime infects popularizations. Every compromise between rigor and simplicity is a bargain with the devil, and I have struck mine as follows. The presentation of Bell's inequality needs no more than some algebra, and is quite rigorous. Understanding Relativity also requires no more than algebraic manipulation, but enough that a purely mathematical account would tax the patience of the average reader. So I have tried to present Relativity pictorially, so far as possible. The figures in the book present the concepts of Relativity accurately, but demand of the reader some skill in interpretation. Pictures of space-time look misleadingly like pictures of space, and the novice must unlearn some of the conventions of normal pictorial representation to avoid being misled. Newcomers should therefore take great care with the pictures in chapter 2: if those are properly understood, the sequel will be easy.

Quantum theory itself has been another matter. Most of the content before chapter 7 can be understood without much discussion of quantum formalism. That formalism itself also uses no more than linear algebra and vector spaces. Interested neophytes can find enough technical detail in any standard introductory text. A particularly nice and accessible presentation of the requisite mathematics is provided in David Albert's *Quantum Mechanics and Experience* (1992, ch. 2).

Just as professional physics scares off the uninitiated, so does professional philosophy. Philosophers have developed many languages of technical analysis which permit concise communication among the cognoscenti but which make amateurs feel like unwelcome guests. But most clear philosophical ideas can be presented intuitively, shorn of the manifold qualifications, appendices, and terminological innovations that grow like weeds in academic soil. I have been very selective in my discussions of the philosophical corpus, usually focusing on a single proposal which illuminates a region of logical space. I do not pretend to comprehensiveness in my review of the philosophical literature, and can only plead for understanding that my decisions reflect a desire for a short, provocative text.

Finally, I feel I should explain the "metaphysical intimations" of my subtitle. Metaphysics has acquired rather a bad reputation in this century, following the insistence of Kant that all metaphysical speculations must be pursued *a priori*. It was not always so. The fount of metaphysics as a philosophical pursuit is the treatise on First Philosophy by Aristotle which has come down to us as the *Metaphysics*. Aristotle was concerned with analysis of what there is into its most generic categories: substance, quality, quantity, etc. I see no reason to believe that Aristotle thought such an examination could not be informed by experience. At its most fundamental level, physics tells us about what there is, about the categories of being. And modern physics tells us that what there is ain't nothing like what we thought there is.

I have used "intimations" rather than "implications" because we still do not know how this story ends. Quantum theory and Relativity have not yet been reconciled, and so we can now at best only guess what picture of the world will prevail. But we do know enough to make some guesses.

This book would not have come to be without help of all sorts. David Albert, Nick Huggett, Martin Jones, Bert Sweet, Paul Teller, and Robert Weingard all devoted their own time and insight reviewing the manuscript and generously shared their views with me. Abner Shimony pushed me to clarify the models in chapter 6, and thereby saved me from repeating some errors in print. Steve Stich expended considerable effort finding the manuscript a home, and always had a word of encouragement. The National Endowment for the Humanities graciously provided financial support in the form of Summer Stipend FT-36726-92 (money = time). And the atmosphere in which the book was completed was lightened by Clio Maudlin, who also improvised some emendations with her feet.

Preface to Second Edition

Publication of the second edition of this tome affords the opportunity, beside typographical corrections, for two more substantial changes. The first is a new derivation of the Relativistic mass increase formula, to be found in chapter 3. The new derivation is somewhat simpler than that in the first edition, and has the advantage of allowing the exact formula to be obtained by means of a few lines of algebra. There are many methods for deriving the formula, but to my knowledge this one is novel. The second is the addition of an Overview of Quantum Mechanics. The overview contains just the bare mathematical bones of the theory, but that is enough to explain how violations of Bell's inequality are implied by the theory. It is hoped that the overview, while not a complete account of quantum theory, helps make this study more self-sufficient.

Beyond providing the chance for small improvements, the issuing of the second edition invites reflection, at some years' remove, on the plan of the original. Perhaps the most vexing question confronting any study of Bell's inequality is how the role of quantum theory ought to be treated. On the one hand, there is little doubt that Bell's inequality, and the experimental observation of violations of that inequality, would never have been discovered if not for the existence of the quantum formalism. On the other hand, the inequality itself is derived without any mention of quantum theory and the violations are matters of plain experimental fact. So the explication and analysis of the importance of Bell's work can in principle proceed without mentioning quantum mechanics at all. Should an account of Bell's inequality emphasize its historical roots in the great mysteries of quantum mechanics or rather sever those ties in the interest of logical clarity?

In composing this book, I chose the second option, playing down the role of quantum theory in favor of pure experimental results. In retrospect, I stand by that decision: the interpretation of quantum theory is trouble-some enough in its own right to overshadow and confuse the relatively straightforward proof of non-locality. But once the main points have been made, the connections between non-locality and the interpretive problems of quantum theory are both intriguing and instructive. In particular, non-locality appears at exactly the point where the "measurement problem" which infects standard quantum theory is resolved. If one resolves the measurement problem by allowing a real physical process of wave collapse, it is the collapse dynamics which manifests the non-locality, and which resists a fully Relativistic formulation. If one resolves the measurement problem by postulating additional variables beside the wave function, it is the dynamics of these variables which manifests the non-locality and which resists a fully Relativistic formulation. The regrettably widespread opinion that there is no real non-locality inherent in the quantum theory is therefore deeply inter-twined with the regrettably widespread opinion that the measurement problem can painlessly be solved without postulating either additional variables or any real collapse process.

Having thrown some rocks at the hornet's nest of the interpretation of quantum theory in this preface, I am obliged now to do more than turn heel and walk away. Although this book is not the place to thrash out those issues, I have thrashed them from time to time in other venues. Some discussions may be found in Maudlin (1995), (1996), (1997), and (1998).

Finally, I must note that although there has been some discussion of Bell's theorem and non-locality in the eight years between the two editions of this book, there had been, to my knowledge, no fundamental change in the basic logic of the situation, and no real progress in reconciling quantum theory and Relativity.

Preface to Third Edition

The impetus for a third edition of *Quantum Non-Locality and Relativity* arises from several different circumstances. The most important is the development of a fully relativistic, precise physical theory that can produce violations of Bell's inequality for experiments at space-like separation. This theory, called flashy Relativistic GRW, was discovered by Roderich Tumulka, and is described in the new chapter 10. Tumulka's theory settles the logical question about the possibility of fully reconciling quantum theory and Relativity. But as the reader will see, the theory also suggests that the consummation was perhaps not so devoutly to be wished. That is, there is a price of plausibility to be paid by this theory, a price so high that it is unlikely to gain converts. One nice thing about flashy GRW is that it forces us to confront deep issues about what exactly makes for a plausible physical theory.

Interest in Bell's inequality has been on the rise in recent years for other unrelated reasons. John Conway and Simon Kochen produced the so-called "Free Will Theorem," whose title alone is enough to raise eyebrows. The theorem utilizes a situation in which there is a violation of Bell's inequality, and Conway and Kochen have made some rather astonishing claims about its significance. This theorem is also discussed in the new chapter.

More generally, interest in Bell's theorem has been reinvigorated by the rise of quantum information theory and the experimental verification of such quantum effects as the teleportation of quantum states. The information-theoretic analysis of quantum theory has had the entirely salubrious effect of focusing attention on the key aspect of quantum theory called *entanglement*. The entanglement of distant systems is what produces violations of Bell's inequality in quantum theory, and physicists have come to routinely accept this entanglement as a quantifiable, exploitable physical resource. There is little dispute any more that the entanglement of distant systems is somehow physically real.

The reason that this development is so cheering is that it deflects attention from other less important aspects of quantum theory. For example, it has been repeated *ad nauseam* that Einstein's main objection to quantum theory was its lack of determinism: Einstein could not abide a God who plays dice. But what annoyed Einstein was not lack of determinism, it was the apparent failure of *locality* in the theory on account of entanglement. Einstein recognized that, given the predictions of quantum theory, only a deterministic theory could eliminate this non-locality, and so he realized that a local theory must be deterministic. But it was the locality that mattered to him, not the determinism. We now understand, due to the work of Bell, that Einstein's quest for a local theory was bound to fail.

Schrödinger, in his famous "cat" paper, remarked on the "*entanglement* of our knowledge of [...] two bodies" (1935, p. 161) found in the quantum-mechanical formalism, but denied, as Einstein did, that this could reflect any real physical connection between separated systems: "Measurements on separated systems cannot directly influence each other – that would be magic" (ibid., p. 164). Bell's work has shown that the magic is real, and physicists who study entanglement have accepted the non-locality that Einstein and Schrödinger could not abide. I have briefly adverted to recent work on the information-theoretic implications of quantum theory in chapter 6 of this new edition.

When I first wrote *Quantum Non-Locality and Relativity*, I tried to keep discussion of the foundations of quantum theory to a minimum. All that is relevant to Bell's theorem are the predictions of quantum theory, not how the theory itself is understood. Separating these issues was especially important at the time because discussions of quantum theory *per se* contained a wealth of distractions and confusion. Perhaps the time is at last ripe to open up the dialog again, and to recover an understanding of Einstein's and Schrödinger's real concerns through the lens of Bell's theorem. I hope that chapter 10 provides a small step in the direction of a clearer understanding of what a comprehensible presentation of any physical theory (and hence a comprehensible presentation of quantum theory) demands.

Introduction

In the 1930s, Otto Neurath was one among many philosophers engaged in the project of purifying scientific language of its ambiguities, its vagueness, and its "metaphysical" contents. One might hope to accomplish this task by an act of radical innovation, building anew from elements of perfect clarity and precision. Neurath realized that such hopes are unattainable, that at best we can only successively improve the language we have, always retaining some of its deficiencies. He illustrated our situation with a resonant image:

> No *tabula rasa* exists. We are like sailors who must rebuild their ship on the open sea, never able to dismantle it in dry-dock and reconstruct it there out of the best materials. (Neurath 1959, p. 201)

The physical sciences themselves suffer the same fate. Fundamental conceptual changes occur, but they are always modifications of a previously existing structure. The entire edifice is not reconstituted anew; rather, tactical adjustments are made in order to render the whole consistent. The *ad hoc* nature of this procedure may leave us with lingering doubts as to whether the whole really is consistent.

During the past century our physical picture of the world has undergone two revolutionary modifications. The Theory of Relativity has overthrown classical presumptions about the structure of space and time. The quantum theory has provided us with intimations of a new conception of physical reality. Classical notions of causality, of actuality, and of the role of the observer in the universe have all come under attack. The ultimate outcome of the revolutions is now but dimly seen, at best. The final reconciliation of

Quantum Non-Locality and Relativity: Metaphysical Intimations of Modern Physics,
Third Edition. Tim Maudlin.
© 2011 Tim Maudlin. Published 2011 by Blackwell Publishing Ltd.

quantum theory and Relativity is a theoretical problem of the first magnitude. No quantum version of General Relativity exists, and the prospects for one are murky. But even apart from that hurdle, problems about the consistency of our two fundamental physical theories may appear.

The problem that will concern us here is easily stated. It arises from the remarkable results derived by John Stewart Bell in 1964 concerning the behavior of certain pairs of particles that are governed by quantum laws. Bell showed that observable correlations between the particles could not be accounted for by any theory which attributes only locally defined physical states to them. The particles appear to remain "connected" or "in communication" no matter how distantly separated they may become. The outcome of experiments performed on one member of the pair appears to depend not just on that member's own intrinsic physical state but also on the result of experiments carried out on its twin.

Many features of this quantum connection are puzzling. It is, for example, entirely undiminished by distance. This distinguishes it from any connection mediated by a classical force, such as gravity or electromagnetism. But even more amazingly, the connection exists even when the observations carried out occupy positions in space and time which cannot be connected by light rays. The particles communicate faster than light.

It is this last feature which raises questions about the consistency of our fundamental theories. Relativity is commonly taken to prohibit anything from traveling faster than light. But if nothing can go faster than light, how can the particles continue to display the requisite correlations even when greatly separated? The two pillars of modern physics seem to contradict one another.

The predicted correlations have been experimentally confirmed. Indeed, they have been seen even in conditions where the communication between the particles would require superluminal velocities. So we are presented with the problem of determining whether Relativity has been violated, and, if so, whether our present account of space-time structure must be modified or abandoned.

The question of whether the quantum correlations are consistent with Relativity seems precise enough to admit a decisive answer, but on closer examination this appearance of clarity dissolves. Exactly what sort of constraints Relativity imposes on physical processes is a matter of much dispute. Many physicists and philosophers would agree that Relativity prohibits *something* from going faster than light but disagree over just what that something is. Among the candidates we may distinguish:

> Matter or energy cannot be transported faster than light.
> Signals cannot be sent faster than light.
> Causal processes cannot propagate faster than light.
> Information cannot be transmitted faster than light.

Most of these prohibitions are easily seen to be non-equivalent. For example, signals could in principle be sent without any accompanying transmission of matter or energy. Or again, superluminal causal processes could exist which, due to their uncontrollability, could not be used to send signals.

Yet another interpretation holds that Relativity requires only that

Theories must be Lorentz invariant.

This requirement is compatible with the violation of every one of the prohibitions listed above.

Not surprisingly, the various prohibitions are justified by different considerations. In one case it is claimed that a violation of the prohibition would require an infinite amount of energy, in another than it would engender paradox, in yet another that some relativity principle would be abrogated. We are therefore left with a rather tangled thicket of problems. We must consider each of the proposed prohibitions and ask whether it is violated by the quantum connection. We must ask how each prohibition is justified and how it connects with the formalism of the Theory of Relativity. We would also like to see how the prohibitions relate to one another. Until this work is done we cannot begin to evaluate the implications of the quantum correlations for our picture of the world.

This problematic directly dictates the structure of our inquiry. Chapter 1 presents Bell's results with a minimum of technical machinery. Chapter 2 is a short intuitive account of Special Relativity. The following four chapters examine the four prohibitions listed above, tracing their connection with Special Relativity on the one hand and their compatibility with quantum non-locality on the other. Chapter 7 delves into the technical requirement of Lorentz invariance and its implications. Chapter 8 touches on the difficulties involved in passing from the space-time of Special Relativity to that of General Relativity.

Any book which attempts to deal with quantum theory, Special Relativity and General Relativity courts various forms of disaster. Technical and mathematical detail can easily push the discussion beyond the ready grasp of the general reader, and the philosophical interpretation of the mathematical formulae can be even more daunting. In this last respect an asymmetry regarding our two fundamental theories should be noted. Relativity is quite well understood. Although it employs ideas that depart radically from those of classical physics, the concepts are themselves unproblematic and become quite transparent with use. Quantum theory, in contrast, still presents deep and basic interpretational problems, the discussion of which could fill several volumes. Fortunately, our concerns will not draw us much into these controversies. Bell's theorem can be proven without so much as a mention of quantum theory, and although one uses quantum theory to predict the violation of Bell's inequalities, the violation itself is confirmed by straightforward laboratory technique. The observed facts, not merely some interpretation of the theory, stand against locality, so the thorny problems surrounding the interpretation of quantum formalism can be almost entirely avoided.[1] For aficionados, more detailed remarks concerning the interpretation of quantum theory will be provided in appendices or in end-of-chapter notes.

Technical details of physics are not the only casualties of our approach. The philosophical literature on this subject is large and growing, and we will be forced to pass over much of it with little examination. I hope that the philosophical views discussed will be accepted by my colleagues as simplifications rather than caricatures.

For those interested in the fundamental structure of the physical world, the experimental verification of violations of Bell's inequality constitutes the most significant event of the past half-century. In some way our basic picture of space, time, and physical reality must change. These results, and the mysteries they engender, should be the common property of all who contemplate with wonder the universe we inhabit. So in telling this tale I have tried to leave behind the arcane technicalia of the academy. In doing so, I have sacrificed no small degree of precision, and perhaps also some important subtleties. But I hope at least to have provided a framework sturdy enough and correct enough to serve both professional and amateur naval architects who propose to redesign the craft which carries us on our journey.

Note

1   To be precise, the only assumption we will be making is that when one does, for example, a polarization experiment and gets some result (photon passed or absorbed), there is, after the experiment is finished, something in the physical state of the universe which picks out that result over the other possible results. Our assumption is held in common by all wave-collapse theories, whether collapse is caused by interaction with macroscopic devices, by conscious experience, or by random "hits" as in the theory of Ghirardi, Rimini, and Weber (1986). It is also held by no-collapse theories such as Bohm's which use additional variables to describe the world. Indeed, I know of only two interpretations which deny the assumption: the many-worlds interpretation of Everett and Wheeler (De Witt and Graham 1973) and the many-minds interpretation of David Albert and Barry Loewer (1988, 1989; Albert 1992). The many-worlds theory is incoherent for reasons which have been often pointed out: since there are no frequencies in the theory there is nothing for the numerical predictions of quantum theory to mean. This fact is often disguised by the choice of fortuitous examples. A typical Schrödinger-cat apparatus is designed to yield a 50 percent probability for each of two results, so the "splitting" of the universe in two seems to correspond to

the probabilities. But the device could equally be designed to yield a 99 percent probability of one result and 1 percent probability of the other. Again the world "splits" in two; wherein lies the difference between this case and the last?

Defenders of the theory sometimes try to alleviate this difficulty by demonstrating that in the long run (in the limit as one repeats experiments an infinite number of times) the quantum probability assigned to branches in which the observed frequencies match the quantum predictions approaches unity. But this is a manifest *petitio principii*. If the connection between frequency and quantum "probability" has not already been made, the fact that the assigned "probability" approaches unity cannot be interpreted as approach to certainty of an outcome. All of the branches in which the observed frequency diverges from the quantum predictions still exist, indeed they are certain to exist. It is not highly likely that I will experience one of the frequencies rather than another, it is rather certain that for each possible frequency some descendants of me (descendants through world-splitting) will see it. And in no sense will "more" of my descendants see the right frequency rather than the wrong one: just the opposite is true. So approach of some number to unity cannot help unless the number already has the right interpretation. It is also hard to see how such limiting cases help us: we never get to one since we always live in the short run. If the short-run case can be solved, the theorems about limits are unnecessary; if they can't be then the theorems are irrelevant.

The many-minds theory does not have this problem, and may be the only existing interpretation of quantum theory which requires no non-local effects. We will discuss the many-minds theory in chapter 7.

1

Bell's Theorem: The Price of Locality

According to our naive, everyday conception, and even according to most of our refined theories, the physical world is composed of separate individually existing objects. The book on my desk sits apart from the glass, each constituted separate from the other and with its own intrinsic properties. The book has its mass, shape, number of pages, the marks of its history engraved on it. It is made up of atoms, each with its own physical constitution, tied together by chemical bonds. The glass similarly exists on its own, constructed from a separate complement of particles. There are, of course, relations between the book and the glass. The book is heavier and occupies more volume; there is a certain definite distance between them. Spatial separation plays a unique role: as an external relation it is not determined by any facts about the book and the glass taken individually. But once we have taken into account their intrinsic properties and their situation in space we appear to have exhausted the facts about the pair. All other facts about them are determined by these.

Each of the pair may influence the other. The glass, full of steaming tea, raises the temperature of the book which is in its proximity. But this interaction is mediated by other localized bits of matter. Air molecules around the glass are made more energetic through interactions with the tea, some wander off and communicate their energy with the book, heating it. The book exerts a slight gravitational pull on the glass and vice versa. This is a subtle matter, but we come to think of this too as a mediated interaction, an effect of a gravitational field.

The fields of classical physics are not so familiar as books or atoms but they too are local entities. Although an electric field may spread out and

Quantum Non-Locality and Relativity: Metaphysical Intimations of Modern Physics,
Third Edition. Tim Maudlin.
© 2011 Tim Maudlin. Published 2011 by Blackwell Publishing Ltd.

permeate the universe, the state of the field is determined entirely by its value at each point of space. Disturbances propagate through the field, but they do so by local interactions: changes in the field quantities induce other changes nearby and so ripple off to infinity. Like the transmission of heat, this process takes time as the vibrations of the field are passed along.

Einstein set great store by the idea that the physical state of the universe is determined by a set of locally defined physical magnitudes so that the state of any localized entity exists independently of all spatially separated systems. As he expressed it in a letter to Max Born:

> If one asks what, irrespective of quantum mechanics, is characteristic of the world of ideas of physics, one is first of all struck by the following: the concepts of physics relate to a real outside world, that is, ideas are established relating to things such as bodies, fields, etc., which claim "real existence" that is independent of the perceiving subject – ideas which, on the other hand, have been brought into as secure a relationship as possible with the sense-data. It is further characteristic of these physical objects that they are thought of as arranged in a space-time continuum. An essential aspect of this arrangement of things in physics is that they lay claim, at a certain time, to an existence independent of one another, provided these objects "are situated in different parts of space". Unless one makes this kind of assumption about the independence of the existence (the "being-thus") of objects which are far apart from one another in space – which stems in the first place from everyday thinking – physical thinking in the familiar sense would not be possible. It is also hard to see any way of formulating and testing the laws of physics unless one makes a clear distinction of this kind. This principle has been carried to extremes in the field theory by localizing the elementary objects on which it is based and which exist independently of each other, as well as the elementary laws which have been postulated for it, in the infinitely small (four-dimensional) elements of space.
>
> The following idea characterizes the relative independence of objects far apart in space (A and B): external influence on A has no direct influence on B; this is known as the "principle of contiguity," which is used consistently in the field theory. If this axiom were to be completely abolished, the idea of the existence of (quasi-) enclosed systems, and thereby the postulation of laws which can be checked empirically in the accepted sense, would become impossible. (Born 1971, pp. 170–1)

Bell's theorem addresses the implications, and ultimately the tenability, of this picture.

Given the extreme generality of the local conception of reality it is hard to imagine that it could, by itself, have any testable empirical consequences. No constraints have been put on the nature or complexity of the locally defined quantities. The locality condition allows, for example, that every particle in the universe could retain traces of every interaction it has ever undergone. It allows a system to be governed by laws which are deterministic or are probabilistic, placing no limit on the subtlety or sophistication of the laws. Nonetheless, Bell was able to show that some behavior of separated pairs of systems cannot be explained by *any* local physical theory if the systems do not interact. Although Bell's results can be derived in different ways and with great generality, we will begin by focusing on a singular fact about light.

Polarization

When one passes a beam of sunlight through a polarized filter, such as the material used in Polaroid sunglasses, two things happen. First, about half of the light is absorbed and half transmitted, as is immediately evident. Second, the light which is transmitted displays an entirely new and surprising characteristic: it shows a particular directionality. This directionality can be most easily observed if one passes the new beam through a second polarized filter. The effect of the second filter depends critically on its orientation with respect to the first. In one orientation the second polarizer will have no effect at all, allowing the entire beam to pass. But as it is rotated, the second filter allows less and less of the light through. By the time it has been turned 90°, it absorbs the beam entirely; as it is rotated further it permits ever more light to pass until, at 180°, the whole beam passes again.

The directionality that the sunlight acquires depends on the orientation of the first polarizer. When the first filter is rotated, the characteristic orientation at which the transmitted beam passes the second filter rotates with it. So light which has passed through a Polaroid filter acquires a new property, a polarization, which is associated with some direction perpendicular to its line of motion.

All that really concerns us is the behavior recounted above; the explanation of the phenomena will ultimately be irrelevant to our concerns. But to help fix our ideas it may help to recall the classical theory of polarization. The classical theory provides us with a simple picture of polarization which should, however, be taken *cum grano sails*, for it cannot be straightforwardly extended when quantum phenomena are taken into consideration.

According to classical physics, light is an electromagnetic wave, a propagating disturbance of the electric and magnetic fields. The fields that vary always point perpendicular to the direction of motion of the light. At any given moment the electric and magnetic fields are also perpendicular to each other, but as time goes on their direction and magnitude may change in any number of ways. For example, if we look at a ray of light head on as it comes toward us, the electric field may rotate in a circle, either clockwise or counterclockwise (circularly polarized light); or it may trace out an ellipse,

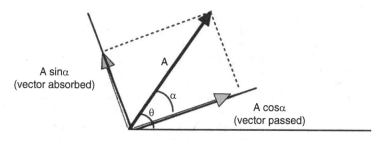

Figure 1.1   Resolving a Vector

rotating and varying in length; or it may simply oscillate back and forth without rotating, always remaining in a plane that points in a given direction. This last possibility, plane polarized light, is the case of interest to us. Plane polarized light has a characteristic direction, the direction of the plane in which the electric field vector always lies. Furthermore, for any direction θ we choose, light of any sort can be analyzed into a component plane polarized in that direction and a component polarized in the perpendicular (that is, θ + 90°) direction. Even circularly polarized light can be constructed from two such elements, if they are added together in the right way, with the right phase relations.

The phenomena recounted above are now easily explained. A Polaroid filter in effect analyzes all incoming light waves into two parts: one plane polarized in the direction of the filter's polarization, the other perpendicular to that direction. It then absorbs the perpendicular component, allowing only the plane polarized remainder to pass through. If the incoming light is unpolarized, this means that on average half of it will pass through and half be absorbed. The effect of the second filter then depends crucially on its orientation relative to the first. If they are perfectly aligned, the light which passes the first is already polarized in the direction of the second and so all gets through. If the second is misaligned by 90°, then exactly the component which passes the first will be absorbed by the second, and none will get through.

What if the two filters are misaligned by some angle α between 0° and 90°? We can represent the light coming through the first filter by a vector pointing in the θ direction whose length A represents the maximum amplitude of the electric field. The second filter resolves this vector into two components, one parallel to θ − α, the other perpendicular (see figure 1.1). The perpendicular component is absorbed by the filter, so the amplitude of the transmitted light is A cos α.

We now must appeal to a seemingly minor but highly significant fact. The energy of plane polarized light is proportional to the *square* of the amplitude of its electric field vector. So if we measure the amount of light

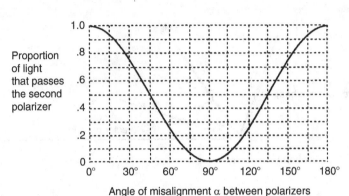

Figure 1.2   Proportion of Light Passing Second Polarizer

which passes the second polarizer by the energy of the beam, the proportion of the beam that gets through is $A^2 \cos^2 \alpha / A^2 = \cos^2\alpha$. Figure 1.2 shows the proportion of the beam which passes the second filter as a function of the angle of misalignment α.

As expected, when $\alpha = 0°$ and the filters are aligned, all the beam is transmitted. When the filters are misaligned by 90° none of the light gets through. But the most significant behavior is found between these extremes. For the moment we need only note that when $\alpha = 30°$, $\cos^2 30° = \left(\sqrt{3}/2\right)^2 = \dfrac{3}{4}$ of the beam gets through, while when $\alpha = 60°$, $\cos^2 60° = \left(\dfrac{1}{2}\right)^2 = \dfrac{1}{4}$ of the light is transmitted.

Light Quanta

According to the classical conception light is a wave, spread out in space. Whenever a plane polarized beam impinges on a filter oriented at, say, 30° off of the polarization plane of the incoming beam the same thing happens: $\frac{3}{4}$ of the beam passes, and what gets through is polarized in the direction of the filter. A beam always comes out with its amplitude and energy reduced by a fixed proportion.

But as Einstein observed in 1905, light does not always behave like a wave. For example, when light falls on certain metals it can knock out electrons causing a current to flow, the so-called photoelectric effect. When one measures the energy of the electrons so liberated one finds that the energy of the incident light is not delivered uniformly over the surface of the metal as one would expect. The energy rather comes in small but discrete packets.

At fine levels of analysis light behaves as if it is made up of particles. These light quanta, or photons, can be individually registered and counted by photomultiplier tubes.

The exact nature of the wave/particle duality of light need not detain us. We need only note two experimentally verifiable facts. First, light from certain sources has the effect of causing discrete, countable events in certain detection equipment. Second, if this light is passed through a polarizer, the resulting beam also behaves as if made up of photons and *the photons are each of exactly the same energy as those in the incoming beam.*

Nothing we have said so far could have prepared us for this new piece of information. It would have been plausible to guess instead that the photons coming through the polarizer would all have had their energy reduced by the same proportion as the energy of the beam as a whole. But in fact the transmitted photons are as energetic as the incoming ones, only the orientation of their polarization is changed by passage.

If the photons which survive the second polarizer each have the same energy as the incoming light quanta, how is the overall energy of the beam reduced? The only possibility is that the light coming out of the polarizer contains *fewer* photons than the light going in. Photons appear to be either transmitted complete through the filter or else swallowed whole.

It is worthwhile to note that all of this talk about light quanta need not be made precise. We could instead refer only to the observable behavior of pieces of laboratory equipment. When light from certain sources is directed at photomultiplier tubes discrete and countable events occur. When the light is passed through a filter fewer such events occur. A second filter again reduces the number, and the proportion of the reduction is the square of the cosine of the angle between the filters. These are the sorts of facts that we will be concerned to explain. The photon picture provides a convenient model of the underlying process, but the correctness of that model need not concern us. If the reader is puzzled by the particulate nature of light it may help to note that experiments similar to the ones we will describe can also be carried out on protons and electrons, archetypical particles. In those cases one measures the so-called "spin" of the particles by passing them through an inhomogeneous magnetic field.

If light behaves as if made up of quanta and if each such quantum which survives a filter has the same energy as it had coming in, then figure 1.2 takes on a new significance. The quantity $\cos^2 \alpha$, which previously measured the proportion of the beam that passes the filter, now represents *the probability for each photon to pass.* If the energy of the beam is to be reduced to one quarter of its previous value when passed through a polarizer oriented at 60° it must be that only one quarter as many photons will compose the passed beam. And if we can create individual photons, it must be that their individual probability for surviving the polarizer is one out of four. As we turn the second polarizer from perfect alignment to perfect misalignment, the likelihood of each photon to get through the second polarizer drops in accord with the graph in figure 1.2.

The Entangled State

So far nothing much mysterious has happened. Polarization phenomena are not particularly strange, and the quantization of light, if unexpected, seems perfectly comprehensible. But one final observation, also apparently rather pedestrian, turns out to be enough to destroy our accustomed picture of physical reality.

When calcium vapor is exposed to lasers tuned to a certain frequency it fluoresces. As excited electrons in the atoms cascade down to their ground state they give off light. In particular, each atom emits a pair of photons which travel off in opposite directions. The polarization of the photons individually shows no preferred direction: for any randomly chosen direction θ the photons will pass a polarizer oriented in that direction half the time. But although the photons individually show no particular polarization, the pairs exhibit some striking correlations. Roughly, each member of a pair always acts as if it has the same polarization as its partner.

More precisely, the following can be observed.[1] Suppose that one photon, R, goes off to the right while its partner, L, goes off to the left. R and L will each eventually impinge on a filter which sits before a photomultiplier tube. If the two filters are set in the same direction then either both photons will pass the filter or both will be absorbed. When the filters are aligned, in whatever direction, the photons are *perfectly correlated*: each does what the other does. If the filters are misaligned, then the photons still behave as if they have the same polarization. That is, suppose R passes through its polarizer, which is oriented in direction θ. Then L will act as if it is polarized in direction θ. If the left polarizer is also oriented in direction θ then L will pass, as we have seen. If the left polarizer is oriented at $\theta + 90°$ then L will be absorbed. And if the angle of misalignment θ is between $0°$ and $90°$ then L will pass the filter a proportion $\cos^2 \alpha$ of the time. Similarly, if R is absorbed by its polarizer, L will act as if it is polarized in the $\theta + 90°$ direction. It will always pass a polarizer oriented at $\theta + 90°$, always be absorbed by one at θ, and generally if the filter is oriented at $\theta + \alpha$ the photon will pass $\sin^2 \alpha$ of the time.

Let us say that a pair of photons *agree* if they are either both passed or both absorbed by their respective filters and *disagree* if one is transmitted while the other is not. Then if the two filters are aligned in the same direction the photons will always agree, half of the pairs being jointly passed, the other

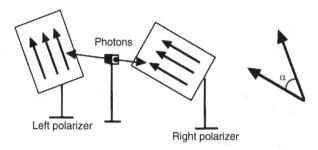

Figure 1.3   Experimental Set-up

half jointly absorbed. If the polarizers are misaligned by 90° the photons always disagree, one being absorbed, the other not. And for any other angle of misalignment α the percentage of pairs which agree (in the long term) is $\cos^2 \alpha$, as shown in figure 1.2.

Note that when we set up a particular experiment we have two choices to make. First we must choose the angle θ of the right-hand polarizer. Then we choose the degree of misalignment α of the left-hand polarizer. If we decide to examine a case of perfect alignment ($\alpha = 0°$) we are still at liberty to set the pair of filters in any direction θ we choose. The fact that we have two free variables, θ and α, is just a reflection of the fact that we have two decisions to make: the angle of the right polarizer and the angle of the left. But no matter how we set the two, the only relevant parameter for calculating the probability of agreement is α, the degree of misalignment (see figure 1.3). If $\alpha = 30°$ the photons will agree $\frac{3}{4}$ of the time; if $\alpha = 60°$ they will agree one time out of four. These simple facts about pairs of photons emitted by calcium vapor are enough to destroy any theory according to which physical reality is local.

How Do They Do It?

Suppose that you and a friend are set the task of reproducing the behavior of the photons: one of you will play photon L, the other photon R. These are the rules of the game: you and your friend start out together in a room (the "calcium atom"). You know that each of you will leave the room by a different door, and after some period of time you will each be asked a question. The question will consist of a number between 0 and 180 written on a piece of paper. Your answer must be either the word "passed" or "absorbed." Before you leave the room, you have no idea which question either of you will be asked. However, while in the room you and your friend are permitted to devise any strategy you please in order to coordinate your answers. Your aim is to ensure that after many repetitions of the game (you are permitted to adopt an entirely new strategy each time) your answers display exactly the same sorts of correlations as the photons show. That is, your strategies must ensure that, in the long run, when the question asked you differs from that asked your friend by an amount α, your answers agree $\cos^2 \alpha$ of the time.

For the moment, we will simplify your task even further. Unlike the photons, which have no information at all about which question will be $\cos^2 \alpha$ asked, you and your friend can know that only one of three possible questions, "0?", "30?" or "60?", will be asked. (We will eventually simplify the task even more, but it is easiest to begin here.) Of course, while you are in the room you still have no idea which of the three questions either of you will be asked. And once you leave the room, we suppose *you have no way of knowing what question has been asked (or will be asked) of your partner*. Your behavior may be determined by your agreed upon strategy and by the question you are asked, but not by the question which your friend happens to be asked. Once again, you must each respond either "Passed" or "Absorbed" when a question is asked.

Over a long run of this game you are aiming to reproduce the behavior of the photons in similar circumstances. That is, after a long series of plays you want to ensure that

Fact 1: When you and your friend happen to be asked the same question you always give the same answer.

Fact 2: When your questions differ by 30, that is, when one is asked "0?" and the other "30?" or one is asked "30?" and the other "60?", you and your friend agree $\frac{3}{4}$ of the time.

Fact 3: When your questions differ by 60, that is, when one of you is asked "0?" and the other "60?", your answers agree $\frac{1}{4}$ of the time.

After all, this is what the photons manage to do.

You and your friend are free to agree on any strategy you like, and you are free to vary your strategy from experiment to experiment. We may suppose that the questions to be asked are chosen at random, so that the pair of the questions "0?" to R and "30?" to L, for example, occurs $\frac{1}{9}$ of the time. It is not, however, important that the questions be asked equal amounts of the time, only that the choices be made at random, so that you can have no idea what is to come. How might you go about settling on a strategy?

The first obvious point is that there is no advantage, and much disadvantage, to using any sort of random element after you have left the room. For suppose your strategy demands that if asked the question "0?" you will decide your answer by a flip of the coin. Since you are unable to communicate with your partner there would be no way for your friend to know how you have answered the question, and so no way to be sure of matching your answer if asked the same question. In general, there is no possible way of satisfying Fact 1 above without deciding in the room *how each of you will answer each question if asked*. For without the knowledge of how your partner would answer a question you cannot act so as to ensure that your answers will match if you happened to be asked the same question.

Besides, no possible advantage can be gained by the introduction of random elements. If one of you may have to flip a coin when asked a question, why not flip it beforehand in the room and share the result with your partner? Or flip it three times, one for each possible contingency. Your partner would then have more information than would be available if you only appeal to the random element when actually asked the question. That excess information cannot possibly *degrade* your performance since, in the worst case, the information can just be ignored. Thus we have the simple result that *any strategy which involves local stochastic elements can do no better than a corresponding strategy where the random choices are made at the source*. A "local stochastic element" is a random process which takes place outside the room and whose outcome cannot be communicated to one's partner. In your case, "at the source" means "in the room"; for the photons it means "in the calcium atom." So in the first place, strategies utilizing local stochastic elements cannot ensure the perfect correlation when identical questions are asked, and in the second place, for every strategy using such elements an equally effective strategy which eschews them exists. If you have a penchant for flipping coins, you may as well flip them in the room. Given these two facts we may now narrow our search to strategies which involve no local stochastic elements. This means that when you leave the room each of you knows exactly what the other will do in each possible situation.

Furthermore, not any such deterministic strategy will do. Since you always run the risk of being asked identical questions, you and your friend must resolve to give the same answer as each other to each question. Only in this way can you assure that when answering identical questions your answers will tally.

So our situation has been greatly simplified. Only eight possible strategies are available, corresponding to the possible ways of answering the three questions. You might, for example, decide to answer "passed" no matter which of the three questions is asked. We will represent that strategy as

⟨P, P, P⟩, where the first slot represents the answer to "0?", the second to "30?" and the third to "60?". The eight possible strategies are then:

(1) ⟨P, P, P⟩   (2) ⟨A, A, A⟩   (A)
(3) ⟨A, P, P⟩   (4) ⟨P, A, A⟩   (B)
(5) ⟨P, A, P⟩   (6) ⟨A, P, A⟩   (C)
(7) ⟨P, P, A⟩   (8) ⟨A, A, P⟩   (D)

Since we are only interested in whether the answers given by you and your friend agree or differ, we can regard each of the corresponding mirror-image strategies above as equivalent. That is, if you choose either strategy (1) or strategy (2) you will agree no matter what pair of questions is asked, if you choose (3) or (4) you will disagree if exactly one person is asked "0?" and agree otherwise, and so on. (Of course there are *other* facts, such as that in the long run approximately half the photons pass and half are absorbed, that would demand a judicious choice between the strategies in the right column and those in the left, but those facts have been omitted from our list.) So we may lump together strategies (1) and (2) calling each "strategy (A)," either (3) or (4) will be "strategy (B)," (5) or (6) "strategy (C)," (7) or (8) "strategy (D)." In order to ensure the strict correlations of Fact 1, you and your friend must choose among strategies (A), (B), (C), and (D) every time a new experiment is run. The only real option that is left open to you, then, is what proportion of the time each strategy will be chosen.

Let us suppose that your decisions over the long run result in choosing strategy (A) a proportion α of the time, strategy (B) β of the time, strategy (C) γ of the time, and strategy (D) δ of the time. α, β, γ, and δ must all be positive numbers (or zero), and of course $\alpha + \beta + \gamma + \delta$ must equal unity.

You and your friend must make your choice of which strategy to adopt in complete ignorance of what questions you are to be asked. Further, we may assume that the choice of questions is determined by a process which is random with respect to your choice of strategy. The experimenters, however they decide which questions to ask, do not do so by predicating their choice on your predetermined strategy. In these circumstances, the long-run results of many repetitions of these experiments will depend solely on the values of α, β, γ, and δ. For example, suppose we wish to know how often the pair of questions "0?", "60?" will receive answers which disagree. They will do so exactly when you have chosen strategy (B) or strategy (D), as can be verified by inspection. In the long run, you choose those strategies $\beta + \delta$ proportion of the time. And since the selection of experiments in which that pair of questions is asked constitutes a random selection from the sequence of strategies you choose, in the long run that pair of questions will receive disagreeing answers $\beta + \delta$ of the time.

By only selecting among the eight strategies we have ensured that Fact 1 will be satisfied. What of the other Facts? Fact 2 states that when the pair of questions "0?" and "30?" or "30?" and "60?" are asked, your answers will agree $\frac{3}{4}$ of the time. Another way of putting this is that your answers will disagree $\frac{1}{4}$ of the time. Similarly, Fact 3 states that when asked the pair of questions "0?" and "60?" you must disagree $\frac{1}{4}$ of the time. We have already seen that the proportion of the "0?" – "60?" experiments which yield disagreeing answers is $\beta + \delta$. By similar reasoning the proportion of "0?" – "30?" experiments which yield disagreement is $\beta + \gamma$, and the proportion of "30?" – "60?" experiments which yield disagreements is $\gamma + \delta$. To recover the correlations of the photons, then, you and your friend must arrange things so that

$$\gamma + \delta = 0.25$$
$$\beta + \gamma = 0.25$$
$$\beta + \delta = 0.75.$$

But now the rub becomes apparent. For on the one hand, the first two equations together imply that $(\beta + \gamma) + (\gamma + \delta) = 0.25 + 0.25 = 0.5$. But on the other hand, $(\beta + \gamma) + (\gamma + \delta) = 2\gamma + (\beta + \delta) = 2\gamma + 0.75$ (by the last equation). These results together imply that $0.5 = 2\gamma + 0.75$ or $2\gamma = -0.25$, so that $\gamma = -0.125$. But γ must be a positive number: it is not among your options to choose strategy (C) –12.5 percent of the time. In sum, there is no possible long-term selection of strategies that you and your friend can adopt which will ensure that your answers will display the same correlations as those of the photons.

Bell's Theorem(s)

The result just obtained can be generalized in many ways, all of which address themselves to variations of the question: given collections of two or more particles and a choice of observations that can be carried out on each, what sorts of constraints on the correlations among results can be derived if the observation carried out on one particle cannot influence the result of observations carried out on the others? We have just seen that in the case of two particles with three possible observations on each particle, if the results when the same experiments are carried out on both wings are perfectly correlated then (proportion of disagreement when experiments 1 and 2 are chosen) + (proportion of disagreements when experiments 2 and 3 are chosen) ≥ (proportion of disagreements when 1 and 3 are chosen). We can abstract from the exact nature of the experiments which are carried out, for in any such case the same reasoning leads to $(\gamma + \delta) + (\beta + \gamma) \geq (\beta + \delta)$.

John Stewart Bell inaugurated this line of investigation in his 1964 paper "On the Einstein–Podolsky–Rosen Paradox" (1987, ch. 2). Bell's result is couched in terms of expectation values, that is, long-term averages for observed quantities, and so takes on a slightly different form from ours. Bell also considers a case of perfect anti-correlation (disagreement when the same quantities are measured) rather than perfect correlation. Bell's result is:

$$|P(a,b) - P(a,c)| \leq 1 + P(b,c),$$

where $P(a, b)$ is the expectation value of the product of the two observed results (the observations always yield the values ±1).

Of course the assumption of perfect correlation or anti-correlation is an idealization relative to actual experimental situations: real laboratory conditions at best allow some approximation of perfect agreement or disagreement. Bell's result was further generalized by Clauser, Horne, Shimony, and Holt (1969) to deal with imperfect correlations. In the case of our polarized photons, it is immediately clear that a small relaxation of the perfect correlation condition would not solve the difficulty. Even if you and your friend are lax enough to allow disagreement to occur 20 percent of the time when the same quantities are measured, the remaining correlations (i.e. 25 percent and 75 percent disagreement in the other experimental set-ups) cannot be recovered. The observed values are so far from the constraints imposed by the locality condition that no small perturbation will bring us back into the allowable range.

Of all the variations on Bell's theorem the most useful pedagogically is that of Greenberger, Horne, and Zeilinger (1989) (GHZ). The GHZ scheme involves three particles rather than two and has the advantage that all of the probabilities involved are either 0 or 1. Since there have been no experimental tests of the GHZ scheme, though, the exposition of it has been relegated to appendix A.

All of the Bell-type results express restrictions on the correlations that can be expected among various experimental observations. The only assumption needed to derive the restrictions is, as we have seen, that the experiment carried out on one particle can have no influence on the outcomes of observations on the other particle. You and your partner can have no advance knowledge of the questions that are to be asked, nor can you later acquire any information about what has been asked your partner so that you could adjust your own answer accordingly. If this condition were violated then the problem could be solved trivially. You need only first agree how your partner will answer each question. Once you know the question which has been asked her, you can decide how to answer the question asked of you. If the same question is asked both of you, you give the matching answer. If the questions differ by 30, you disagree 25 percent of the time; if they differ by 60, you disagree 75 percent of the time. Only information about the question asked need be transmitted since you could agree beforehand how your partner will respond.

A final remark on Bell's theorem is in order. Bell himself derived the result as part of an examination of so-called local hidden-variables theories. Such theories attempt to eliminate the stochastic element of orthodox quantum theory by adding extra parameters to the usual quantum formalism, parameters whose values determine the results of the experiments. Bell's results are therefore sometimes portrayed as a proof that local deterministic hidden-variables theories are not possible.

This is a misleading claim. It suggests that the violation of the inequalities may be recovered if one just gives up determinism or hidden variables. But as we have seen, the only assumption needed to derive the inequalities is that the result of observing one particle is unaffected by the experiment carried out on the other. Subject to this restraint, no deterministic or stochastic theory can give the right predictions, no matter how many or how few variables are invoked.

Since a natural method of attempting to ensure the isolation of the two particles from one another is to carry out the relevant observations in distantly separated places, the isolation condition is generally called "locality." The assumption involved is that observations made on one photon can in no way alter the dispositions of the other photon to pass or be absorbed by its polarizer. Adopting this terminology uncritically for a moment, we have shown that Bell's inequality must be obeyed by any local theory of any sort. Adding stochastic elements does not help the situation at all, as noted above. So experiments verifying the violation of Bell's inequality would doom locality *tout court*.

Aspect's Experiment

To carry out exactly the experiment outlined above we would construct two polarization analyzers which could each be quickly set to any of three different positions: 0°, 30° or 60°. On reflection, though, we can see that the situation can be simplified even further. Suppose that each analyzer can be set at only one of two positions. We could then arrange things so that the right polarizer could be set at either 0° or 30° while the left can be set at 30° or 60°. The statistics to be reproduced are the ones we have already derived: when both analyzers are set at 30° the photons always agree; when we have 0° on the right and 30° on the left or 30° on the right and 60° on the left they agree 75 percent of the time; when the right is set at 0° and the left at 60° they agree only 25 percent of the time. The analysis of this situation goes just as before, with exactly the same strategies available. Strategy ⟨P, A, A⟩ represents the decision that the right-hand photon will pass if measured at 0°, both will be absorbed if measured at 30°, and the left-hand photon will be absorbed if measured at 60°.

Indeed, the impossibility of satisfying these conditions is even more obvious now. The photons must agree on how they will both act if measured at 30°. In order to achieve the 75 percent agreement rate for the 0°–30° possibility, 25 percent of the time the right-hand photon must choose a strategy in which 0° differs from 30°. To achieve the 75 percent agreement for 30°–60°, the left-hand photon can only allow its 60° value to deviate from the common 30° value 25 percent of the time. But if the right-hand photon lets the 0° value deviate from the 30° value only 25 percent of the time, and the left-hand photon only allows 60° to deviate from 30° 25 percent of the time, then at least 50 percent of the time neither will so deviate. But then at least 50 percent of the time 0° and 60° will agree with each other (since neither deviates from the common 30° value) and the observed 0°–60° disagreement rate of 75 percent cannot be recovered.

The analysis again depends on the primary assumption: the setting of the polarizer on one side cannot be communicated to or have an effect on the photon on the other side. Experimentally we can try to ensure this condition in two ways. First, we want to separate the two analyzers from one another in space. Second, we want to choose the setting of the polarizer at the last possible moment. If the setting is chosen just before the measurement is made then the second photon could not adjust its strategy on the basis of prior knowledge of the question to be asked its partner. That is, if while you and your partner are in the room it has not yet even been decided which questions will be asked, then you cannot agree on a successful strategy while still in the room. Hence an ideal experimental condition will have two polarizers each of which can be set to one of two settings, well separated in space, with quick choices of the settings being made.

Such an experimental situation was realized in 1982 by Alain Aspect and his collaborators (Aspect *et al.* 1982). Since physically rotating a polarizing filter cannot be done quickly, Aspect hit on a clever means of choosing between the two possible experiments on each side. Two polarizers and detectors were set up on each side of the experiment, with a very fast optical switch which could send the photons to either one (figure 1.4). Each of the optical switches alternated the beam between the two polarizers every 10^{-8} seconds. The right-hand apparatus was about 12 meters away from the left-hand one.[2] (These details will be of some importance in the coming chapters.) For the moment we can only note that it is in no way obvious how the result on the right-hand side could depend on which detector the left-hand photon is sent to. And if no such dependence exists then Bell's inequality cannot be violated: the quantum correlations could not be reliably

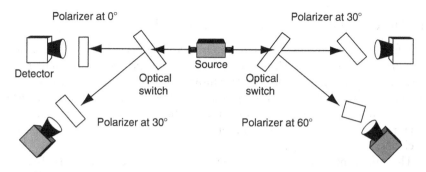

Figure 1.4 Aspect's Experiment

produced.[3] In Aspect's experiment the quantum predictions were confirmed and Bell's inequality was violated.

What Is Weird About the Quantum Connection?

Aspect's experiment and other such experiments have produced observable data which cannot be predicted by any theory which disallows influence of the career of one particle on the behavior of the other once they separate. Somehow the particles must remain in communication, the observable behavior of one being determined, in part, by the nature of the observations carried out on its twin. After being created together the pair of particles remain interconnected.

This interaction among distantly separated particles presents profound interpretive difficulties. But one might initially be surprised that this behavior should elicit any concern at all. After all, classical physics is shot through with such causal connections among distant particles. Newtonian gravitational theory, for example, postulates that every massive particle in the universe exerts a gravitational force on every other, a force of magnitude Gm_1m_2/r^2. When a sparrow falls in Yugoslavia it has effects in New Brunswick and on Saturn and in the most distant galaxy. Some small gravitational tug will register in the smallest parts of the most far-flung stars. In the face of this sort of interconnectedness the quantum connection looks rather modest.

But there are at least three features of the quantum connection which deserve our close attention. All of them are, to some extent, surprising. The first two prevent our assimilation of these quantum effects to those of a force like gravitation. The last presents problems for reconciling the results of experiments like that of Aspect with the rest of our physical picture.

1. The quantum connection is unattenuated

The fall of a sparrow in Yugoslavia may have its effects in New Brunswick and on Saturn and beyond, but the effect becomes progressively smaller the farther away one goes. Since the gravitational force drops off as the square of the distance it eventually becomes negligible if one is concerned with observable effects. The gravitational pull of the sparrow plays no noticeable role in affairs in New Brunswick, much less in the affairs of extra-galactic societies.

The quantum connection, in contrast, appears to be unaffected by distance. Quantum theory predicts that exactly the same correlations will continue unchanged no matter how far apart the two wings of the experiment are. If Aspect had put one wing of his experiment on the moon he would have obtained precisely the same results. No classical force displays this behavior.

2. The quantum connection is discriminating

When the sparrow falls in Yugoslavia I feel a slight gravitational tug in New Brunswick. So does the computer on my desk, and the cat asleep on the bed. Every inhabitant of Princeton is jostled slightly, and to nearly the same extent as the population here. The effects of the sparrow's fall ripple outward, diminishing as distance increases, jiggling every massive object in its way. Equally massive objects situated the same distance from the sparrow feel identical tugs. Gravitational forces affect similarly situated objects in the same way.

The quantum connection, however, is a private arrangement between our two photons. When one is measured its twin is affected, but no other particle in the universe need be. If we create a thousand such correlated pairs and send the right-hand members all off in a group, each particle still retains its proprietary connection with its partner. A measurement carried out on one member of the right-moving hoard will influence only one member of the left-moving group, one particle situated in the midst of a thousand seemingly identical comrades.

The quantum connection depends on history. Only particles which have interacted with each other in the past seem to retain this power of private communication. No classical force exhibits this kind of exclusivity.

3. The quantum connection is faster than light (Instantaneous)

Of all the peculiarities of the particle communication, this might seem to be the most benign. For although no classical forces are unattenuated or discriminating, all were at least originally described as instantaneous. Classical gravitational and electrical forces were described as being determined by the contemporaneous global distributions of matter or of electric charge. Any change in that global distribution would therefore immediately have effects on the forces felt everywhere.

Although instantaneousness was a feature of the first theories of gravitation and electricity, it was not an essential feature. Newton thought that gravitation must be the effect of some subtle particles, about which he famously framed no hypotheses. He would therefore have expected a perfected theory of gravitation to take the speed of these particles into account. In such a final theory one would expect some delay to intervene between the sparrow's fall and the slight jostle it causes in New Brunswick. Of course, in the classical regime no *a priori* constraint could be put on the velocity of the gravitational disturbance, but one might reasonably expect it not to be infinite.[4]

But the modern theory of space and time differs radically from the classical view. The revolution has come in two stages, both initiated by Einstein: the Special and General Theories of Relativity. The Special Theory confers upon light, or rather upon the speed of light in a vacuum, a unique role in the space-time structure. It is often said that this speed constitutes an absolute physical limit which cannot be broached. If so, then no relativistic theory can permit instantaneous effects or causal processes. We must therefore regard with grave suspicion anything thought to outpace light.

The quantum connection appears to violate this fundamental law. Aspect's experiment was so contrived that the setting of the equipment at one side could not be communicated, even by light, in time to influence the other side. All three of these weird aspects of the quantum connection are related to spatial structure. Classical forces depend on spatial separation while the quantum connection does not. The effects of classical forces are determined by spatial dispositions: two electrons near one another will be (nearly) identically affected by distant gravitational or electrical sources. The quantum connection discriminates even among identical sorts of particles which are in close proximity to one another. Finally, the speed of the quantum communication appears to be incompatible with relativistic space-time structure.

Our concern will almost entirely be with the last of these three features. It is surprising that the communication between particles is unattenuated and discriminating, but often our best counsel is simply to accept the surprising things our theories tell us. The speed of the communication is another matter. We cannot simply accept the pronouncements of our best theories, no matter how strange, if those pronouncements contradict each other. The two foundation stones of modern physics, Relativity and quantum theory, appear to be telling us quite different things about the world. To understand, and perhaps resolve, that conflict we must consider carefully just what Relativity tells us about space and time.

Appendix A: The GHZ Scheme

If both you and your partner are uninformed about the question being asked the other, there is no strategy for answering questions which will reliably reproduce the quantum correlations in the long run. But this trouble matching the behavior of photons only appears in the long run: in every individual "experiment" you and your partner can be assured that your responses *in that particular experiment* are responses which the photons might have given. If, for example you decide during one particular game that both players will answer "passed" no matter which question is asked, you can be assured, no matter which questions are asked, that your responses will not in themselves violate any quantum-mechanical predictions. For the only ironclad constraint imposed by the quantum correlations is that both partners give the same answer if they are both asked the same question. So long as you have agreed to a common response to each question you are safe on that run: the answers you give will certainly be quantum-mechanically permissible. Only after many games will your failure to match the target correlations emerge.

This state of affairs is frustratingly equivocal. Non-communicating partners can be certain that in no particular game will they diverge from what the photons might do, but can be equally certain that over time they must diverge in their cumulative behavior. There is something a bit ephemeral or ghostly about the problem, in that it lies entirely in long-term averages. Is this an indication of something deep about the nature of the quantum predictions?

A discovery by Daniel Greenberger, Michael Horne, and Anton Zeilinger (1989) dispels this suspicion. Greenberger, Horne, and Zeilinger (GHZ) found that in some instances quantum theory makes predictions about correlations between particles which are so strong that no local (i.e. non-communicating) strategy can be assured of matching the quantum predictions on any single run of the experiment. Although the GHZ scheme has not been tested in any actual experiment, it merits our attention as an indication of the strongly non-local character of quantum mechanics.[5]

The GHZ scheme uses three particles rather than two, and measures spin rather than polarization. The three particles can be created and allowed to separate to arbitrary distance, at which time a spin measurement is made on each. As in the Bell case, we can model the situation as a game. In this one, you and two partners begin together in a room. Some time after you depart the room, traveling in different directions, each of you will be asked one of two questions. We will denominate the questions "X?" and "Y?"; they correspond to measuring the spin of the particles in either of two orthogonal directions. To each question you must answer either "up" or "down" (in the literature, the responses of the particles are also often represented as 1 and −1). Since each particle can be asked one of two questions, there are eight distinct possible experimental arrangements, but of these only four will ever be used. Either two of the players will be asked "Y?" and the last one "X?" or all the players

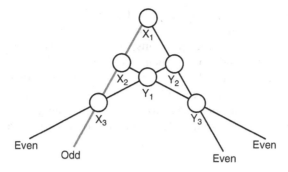

Figure A.1 The GHZ Problem

will be asked "X?". Using an obvious notation, we will represent these four experimental arrangements as $X_1Y_2Y_3$, $Y_1X_2Y_3$, $Y_1Y_2X_3$, and $X_1X_2X_3$.

GHZ noted that for one particular quantum state of a triple of particles, the following predictions can be made *with certainty*. If any of the first three experimental arrangements is chosen then the particles will, among them, respond "up" an *even* number of times. But if X measurements are made on all of the particles, they will, among them, respond "up" an *odd* number of times.

The quantum state does *not* fix exactly which particle will respond "up" and which "down" in any case. Nor does it predict whether the odd number of "up"s will be 1 or 3, or the even number 0 or 2. But according to quantum theory, the responses to the first three experimental arrangements will always include an even number of "up"s and the response to the fourth an odd number. Can you and your colleagues manage to duplicate this feat if each of you remains ignorant of the questions asked the others?

First, it is obvious that only a deterministic strategy will do. In any given experimental arrangement, once two of the partners have answered, the last partner will have to give a particular response in order for the number of "up" responses to come out right. If anyone is using stochastic mechanisms (and is unable to report the result to the other players) then he runs the risk of giving an unacceptable answer.

While in the room, then, you must decide on how each partner will answer each question. The problem situation is vividly illustrated by the graph in figure A.1. Your task, in effect, is to write either "up" or "down" in each of the empty circles in such a way that an even number of "up"s lie in the triples linked by solid lines while an odd number lie in the remaining triple.

But this task is impossible. To solve it, the set of answers to $X_1Y_2Y_3$ must contain an even number of "up"s, as must the answers to $Y_1X_2Y_3$ and $Y_1Y_2Y_3$, while $X_1X_2X_3$ contains an odd number. Adding these up, $X_1Y_2Y_3$ + $Y_1X_2Y_3$ + $Y_1Y_2X_3$ + $X_1X_2X_3$ = even + even + even + odd = odd. But $X_1Y_2Y_3$ + $Y_1X_2Y_3$ + $Y_1Y_2X_3$ + $X_1X_2X_3$ = $2(X_1X_2X_3Y_1Y_2Y_3)$, that is, in taking the sum, the answer to each possible question was counted *twice*. The numbers of "up"s in the sum, then, must be even, not odd.

Since there is no solution to the task, the GHZ scheme brings home the problem for locality all the more sharply. No matter how you and your friends decide to answer the questions, there will always be at least one experimental arrangement for which you are bound to give the wrong answer. If each experimental arrangement is chosen with equal probability, then on average you must give a set of responses which is quantum mechanically *forbidden* at least one time in four. Every time the game is played you must run a risk of failing to do what the real particles are certain to do.

Notes

1   When I write that this behavior can be observed, I am making certain idealizations about detector efficiencies, which do not approach 100 percent. What is correct to say is, first, that quantum mechanics predicts these correlations if the detectors were perfectly efficient. Second, what is observed is exactly in accord with the quantum mechanical predictions if the actual detector efficiencies are taken into account in the usual ways.

2   A very useful review of Aspect's experiment, as well as other experimental tests of violations of Bell's inequality can be found in Redhead 1987, pp. 107ff.

3   The qualifier "reliably" refers to the fact that the photons could, as it were, guess what the polarizer settings on each side will be, adjusting their strategies accordingly, and their guesses *could*, by pure chance, be right. But absent of any causal or informational connection, such accurate guessing would be a miraculous coincidence. The chance of a local system violating the Bell inequalities becomes arbitrarily small as the experiment collects more data.

4   It is interesting to note that such a perfected theory of Newtonian gravitation which incorporates a finite speed for the propagation of gravitational effects would necessarily lead to slightly different predictions than Newton's theory. Paul Gerber used the assumption that gravitational influence travels at the speed of light to derive a corrected equation for planetary orbits. In this entirely Newtonian milieu, Gerber derived exactly the equation which Einstein recovered from the General Theory of Relativity. In particular, Gerber showed that this theory predicts the anomalous advance of the perihelion of Mercury. This is particularly remarkable since Gerber derived his result 17 years *before* Einstein, before the discovery even of the Special Theory of Relativity. Gerber's result is derived in a somewhat more perspicuous way by Petr Beckmann (1987, pp. 170–5). Robert Weingard (p. c.) has confirmed Beckmann's derivation, but remarks that the delayed-propagation version of Newtonian gravitation does not give the General Relativistic prediction for bending of light. Gerber's work was unfortunately seized upon by rabid anti-Relativists and has fallen into disrepute; the notion of a delayed-propagation Newtonian gravitational theory, though, is natural enough to warrant study.

5   For an extensive, but somewhat technical, discussion of GHZ-type schemes, see Clifton, Redhead, and Butterfield (1991a). But see also Jones (1991) and Clifton, Redhead, and Butterfield (1991b).

2

Relativity and Space-time Structure

In the last chapter we saw that no collection of systems, no matter how constructed or governed by law, could reliably produce violations of the Bell inequalities provided that one condition be satisfied: the question asked or experiment done on one side of the apparatus can have no influence on the result at the other side. We denominated that condition "locality," a decision which demands some justification.

If we want to isolate the two wings of the experiment from one another, a natural thing to do is to move far apart. In the classical regime this tactic seems reasonable for several reasons. Since the known classical forces decay with distance one might hope, first of all, to reduce any effects to a negligible magnitude. Further, if the causal connection is mediated by a process which takes time, e.g. by particles or waves which travel at a finite speed, then at a sufficiently great distance the setting of one detector cannot be communicated to the other wing in time to influence its outcome. The further apart the two wings are, the longer messages would take to get from one side to the other.

In the classical regime these considerations yield at best plausibility arguments, not proofs. Gravitation and electrical forces may decay with distance, but nothing in the classical world-view requires this of all forces. And no fixed upper bound exists for the speed of propagation of classical influences. Although spatial separation may suggest causal isolation to the classical mind, it is only suggestion, not implication.

In the relativistic regime the situation looks rather different. The speed of light acquires a central role, serving in at least some contexts as an absolute physical limit. If it is a limit on all causal influences then the inference

Quantum Non-Locality and Relativity: Metaphysical Intimations of Modern Physics,
Third Edition. Tim Maudlin.

from "spatially separated" to "causally isolated" will become, in appropriate circumstances, valid. Our choice of "locality" as a name for the crucial isolation condition would then be vindicated: Einstein's systems A and B, if sufficiently far apart, have their own physical states and "external influence on A has no direct influence on B." Given a finite limit to the speed of causal processes, nor could external influences on A have an indirect (mediated) influence on B.

Our main order of business, then, is to determine whether Relativity really does forbid influences between sufficiently separated events. If so, then Aspect's experiments constitute both a vindication of quantum theory and a refutation of the Theory of Relativity. But even if Special Relativity does not strictly forbid superluminal processes, the situation is very different from that of classical physics. Due to the role of the speed of light in Relativity, superluminal processes must at the least have very exotic properties, signaling a radical change from the more familiar mechanisms of nature.[1]

Our goal in this chapter is to present the Special Theory of Relativity and its account of space-time structure. But as a propaedeutic to this task we begin with a review of some familiar facts about the maps we draw on space and time.

Coordinate Systems: Euclidean Space

We begin by considering the standard two-dimensional Euclidean plane. The plane contains an infinite multitude of points, so in order to describe figures in the plane we need some method for assigning names to the points. We could, in principle, simply invent names for various points, such as "the Grange" or "Chesney Wold," but the inconveniences of such a technique are patent. Arbitrary names convey no information about the relative locations of points, about their distances apart or the directions between them.

The solution to this problem lies in using numbers as names for the points and in assigning the names in a systematic way. At a *minimum* we would like the numbers to be assigned to the points in such a way that nearby points are assigned nearby numbers. More exactly, we would like to arrange things so that as we travel along a continuous trajectory from any point P to any other point Q the numbers assigned to the points we visit vary continuously from those assigned to P to those assigned to Q.

In a two-dimensional plane the only way to satisfy this continuity constraint is to assign (at least) a pair of numbers as coordinates to each point.[2] To coordinatize the plane we lay down two families of curves, the members of each family being parameterized by a real number. Let's call the first family of curves the x-family and the second the y-family. If we choose our families judiciously every point on the plane will lie at the unique intersection

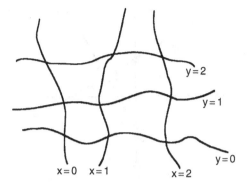

Figure 2.1   A Coordinate System

of one member of the x-family with one member of the y-family. Thus each point gets assigned a unique pair of numbers (x, y), its x-coordinate and y-coordinate.

There are infinitely many ways of laying down families of coordinate curves on the Euclidean plane. One such possibility is depicted in figure 2.1. The coordinate system of figure 2.1 does satisfy the continuity constraint, and so is significantly more convenient than assigning a random proper name to each point. When we get to the space-times of General Relativity in chapter 8 we will be quite satisfied with a system like that of figure 2.1. But given that we now have the Euclidean plane, curvilinear systems like the one depicted have several drawbacks. The amount of information we can infer given only the coordinates of two points is quite limited, since the curves in each family are so dissimilar from one another. We cannot infer, for example, how far the point (0, 0) is from (1, 2), or in exactly what direction it lies. Clearly we can do better than this.

First, it is convenient to use *rectilinear* coordinate curves rather than curvilinear ones so we know that so long as we keep to one such curve we will travel in a straight line. Since no two members of a family can intersect each other, this means we must choose families of parallel straight lines. Parallel lines always retain a constant distance from one another, so it will be helpful to have the numerical values assigned to the members of each family reflect the distance between them. Finally, calculations are simplified if we require the two families to lie at right angles to each other so we can use the Pythagorean theorem to compute distances. In this way we finally arrive at the familiar system of Cartesian coordinates, a rectangular grid.

We have gone such a long way round to a familiar goal to emphasize that Cartesian coordinates are just one of an infinite number of ways to assign numbers to the points in the plane. Further, in order for a Cartesian system to be possible at all, the surface being coordinatized must have several

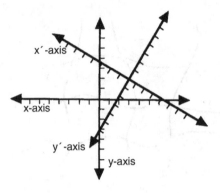

Figure 2.2   Two Cartesian Systems

special symmetries. Such a system cannot, for example, cover the whole surface of a sphere, or any finite region of it.

Even having settled on a rectangular, rectilinear coordinate system we are left with several choices to make. Many distinct Cartesian grids can be put down on a Euclidean plane. Two such systems are depicted in figure 2.2. The x–y coordinates differ from the x′–y′ coordinates in two ways: they have different origins and are oriented in different directions. As a result, every point is assigned a different set of coordinates in each system.[3] But there are equations which allow us to determine the coordinates of a point in one system given its coordinates in the other. For the example depicted in figure 2.2 the transformation equations are:

$$x = \frac{\sqrt{3}}{2}x' + \frac{1}{2}y' + 3$$

$$y = -\frac{1}{2}x' + \frac{\sqrt{3}}{2}y' + 2.$$

Thus the point (0, 0) in the x′–y′ system is called (3, 2) in the unprimed system and the point (3, 4) in the primed system is (7.598, 3.964) in the unprimed.

Let's simplify matters yet further by considering only Cartesian co-ordinates which share the same origin. In this case the systems can differ only in the orientation of their axes (figure 2.3). If the angle between the x and x′ axes is θ then the transformation equations are:

$$x' = x \cos \theta + y \sin \theta$$
$$y' = -x \sin \theta + y \cos \theta.$$

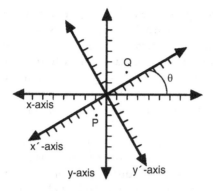

Figure 2.3 Rotated Coordinate Systems

Invariant Quantities

For the moment, consider two Cartesian coordinate systems which share an origin but whose axes are inclined at 30° to one another. The coordinate transformation becomes:

$$x' = \frac{\sqrt{3}}{2}x + \frac{1}{2}y$$

$$y' = -\frac{1}{2}x + \frac{\sqrt{3}}{2}y.$$

The point P in figure 2.3 would then be assigned coordinates $(-1, -2)$ in the x–y system and $(-(\sqrt{3}/2+1), \frac{1}{2}-\sqrt{3})$ in the x'–y' system. Similarly, the point Q is called $(2, 2)$ in the unprimed coordinates and $(\sqrt{3}+1, \sqrt{3}-1)$ in the primed. There are few things on which the two coordinate systems agree. Except for the origin, they assign different coordinates to all the points in the plane. Nor do they agree on coordinate differences. For example, the difference in x-coordinates between Q and P, Δx_{QP}, is $2 - (-1) = 3$ in the x–y system but $\sqrt{3}+1-(-(\sqrt{3}/2+1)) = 3\sqrt{3}/2+2$ in the x'–y' frame. $\Delta y_{QP} = 4$ while $\Delta y'_{QP} = 2\sqrt{3} - \frac{3}{2}$. Despite these differences, certain quantities remain unchanged when we pass from one coordinate system into the other. In particular, although both Δx_{qp} and Δy_{QP} change, the quantity $\sqrt{(\Delta x_{qp}^2 + \Delta y_{qp}^2)} = 5$ remains the same. This is, of course, just the distance between the points as calculated using the Pythagorean theorem.

We can now make a fundamental distinction between two sorts of quantities. *Frame dependent* quantities such a Δx and Δy change from reference frame to reference frame. *Invariant* quantities, such as the distance between two points, are the same in all coordinate systems. The frame dependent quantities are, in part, artifacts of convention. The vector from P to Q does not intrinsically divide itself into two components Δx and Δy. The decomposition effected by the x–y frame is no more natural or correct than that effected by the x'–y' frame. But the distance between the points is not in this way a product of our conventions. Such invariant quantities represent the objective, frame-independent features of points on the Euclidean plane.

This fundamental distinction will be of paramount importance in our investigation. The objective features of the world must be represented by invariant quantities. If many different reference frames are all equally valid, then the only statements which can be held as objective descriptions of the world are statements which are agreed upon by all the frames. In the case of the Euclidean plane the fundamental invariant is distance: the distance relations between the points determine all of the intrinsic features of the plane.

Classical Space-times

From the humble Euclidean plane, we want to advance to the problem of introducing coordinates for both space and time. But first let's consider the small complications involved in moving to spaces of a higher dimensionality than two.

To coordinatize three-dimensional Euclidean space with Cartesian coordinates we introduce three families of mutually orthogonal Euclidean planes, the x-planes, y-planes and z-planes. Once again the families are so chosen that each point lies at the unique intersection of three planes, one from each family. These provide the numerical coordinates of the point. In general, any N-dimensional space is coordinatized by choosing N families of (N − 1)-dimensional spaces so that each point acquires a unique set of coordinates in this way. In a space of N dimensions an (N − 1) dimensional subspace is called a *hyperplane*. So coordinates involve finding N different ways of splitting up the space into collections of hyperplanes.

When we locate an *event* in the world we need to specify four coordinates: three to give the location and one to give the time. Providing these numbers for all events amounts to the problem of laying down coordinates on a four-dimensional space. The properties of this space-time may well, however, differ significantly from those of the corresponding four-dimensional Euclidean space. Constraints on the coordinatization of space-time implicitly contain theories of the objective space-time structure, so we will proceed slowly and with some care. Since a four-dimensional object is hard to imagine we will illustrate the points using two- or three-dimensional space-times.

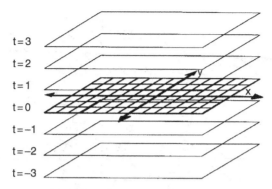

Figure 2.4 Hyperplanes of Simultaneity with Spatial Coordinates on t = 0

The suppression of one or two spatial dimensions is for convenience alone, and the analysis of the full four-dimensional space-time should be clear.

Consider, then, a three-dimensional space-time. In three-dimensional Euclidean space there is no basis for picking out a preferred way of slicing the space into stacks of two-dimensional Euclidean planes. As in the analogous case of figure 2.3, we could orient the axes of the reference system in any way at all. But according to classical physics, space-time naturally falls apart into hyperplane slices in at least one respect. For we certainly must assign the same time coordinate to all of the events which occur at the same time. All classical coordinatizations of space-time therefore agree on one of their families of hyperplanes: the hyperplanes consisting of simultaneous events. Numerical values will be assigned to these hyperplanes which reflect the time elapsed between them. So one of our coordinates has been fixed (up to an arbitrary choice of origin and time scale).

If we choose one of the hyperplanes of simultaneity, say t = 0, the problem of introducing spatial coordinates on it is just the problem of putting down Cartesian coordinates on a Euclidean space. As we have seen, there will be some arbitrariness in the choice of origin and orientation of such a frame. Having made the arbitrary choice we arrive at the situation depicted in figure 2.4. Every event has been assigned a time coordinate, and every event on the surface t = 0 has also been assigned spatial coordinates. So we are left with the problem of *extending* the spatial coordinates from t = 0 to the rest of the space-time.

What sorts of constraints should govern this extension? At a minimum, we want the continuity constraint: the points along continuous trajectories in space-time ought to have continuously varying coordinates. We would also like a Cartesian system of spatial coordinates to be induced on every hypersurface of simultaneity. If these are the *only* constraints that our coordinate frame must meet, then we have a very wide range of choices. One

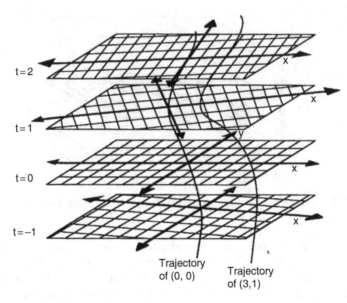

Figure 2.5   A Coordinatization of Space-time

such choice is depicted in figure 2.5. The implicit claim of such liberality in choice of coordinate systems is that the objective structure of space-time consists solely in the absolute division into simultaneity slices plus the spatial geometrical structure on each of those slices (plus continuity). This very weak objective structure is called *Leibnizian space-time*. It does not contain enough structure for classical physics.[4]

Consider the t axis, the trajectory of the point x = 0, y = 0 through time. In Leibnizian space-time, any trajectory which is always future-directed and which passes through the point (0, 0, 0) can serve as the time axis of a coordinate system (by convention, we will take the first coordinate to be time). This is in striking contrast to our coordinatization of three dimensional Euclidean space. In Cartesian coordinates the z axis in such a space must be a straight line orthogonal to the z = 0 plane. Cartesian coordinates are rectilinear and rectangular, and this greatly reduces the reference frames available to us.

So why not simply demand that the t axis of our reference system be a line which is both straight and orthogonal to the t = 0 hyperplane? Here we must proceed with extreme caution. Given our graphical presentation of the three-dimensional space-time it is easy to presuppose that there must be a distinction between curved and straight trajectories through space-time. But does such a distinction really exist? Are there any physical facts which would allow us to distinguish a straight from a curved trajectory, or to calculate the "angle" between such a trajectory and the t = 0 hyperplane? Each of these notions, that of a straight trajectory and that of an orthogonal trajectory, carry along substantive assumptions and cannot be admitted as legitimate without scrutiny.

Does nature make a distinction between straight and curved paths through space-time? The centerpiece of classical physics can be construed as answering in the affirmative. Newton's first Law of Motion reads:

Every body continues in its state of rest, or of uniform motion in a right line, unless it is compelled to change that state by forces impressed on it.

In order for Newton's Law to *make sense*, for it to *make any claim at all*, there must be a distinction in nature between the trajectories of particles which are at rest or in uniform motion and those which are not. So there must be enough objective structure in space-time to found such a distinction.

Our first conclusion, then, is that there *is* a distinction in nature between uniform and accelerated motions. We further postulate that uniform motions trace out straight trajectories through space-time and accelerated motions follow curved paths. (This identification of uniform motion with straight trajectories is most natural since any two bodies each in uniform motion do not accelerate relative to one another.) Newton's first Law simply states that any body not subject to a force travels along a straight path through space-time. A body subject to a force will occupy a curved trajectory, and Newton's second Law states exactly how the path will curve. *Acceleration* of a body is nothing more than the curvature of its trajectory through space-time.

A corollary of Newton's first Law, then, is that we can legitimately demand that the time axis of our coordinate system be a straight line in space-time. Indeed, we can demand that every spatial coordinate sweep out a straight trajectory. Coordinate systems which satisfy this constraint are precisely the inertial frames of classical physics. The postulation of global inertial frames is a bold and risky hypothesis. There is no guarantee that nature has been so kind as to provide us with a space-time which can be coordinatized so that (1) the spatial coordinates trace out straight trajectories and (2) the spatial coordinates maintain the same distances from one another through time. It is a fundamental postulate of classical physics that such global inertial frames exist.

This leaves us still with the question of the orthogonality of the time axis. Does anything in nature distinguish a trajectory which is at right angles to a simultaneity hyperplane from one which is inclined at some other angle?

Again, Newton's first Law provides his answer. The first Law mentions a distinction between bodies in a state of rest and those in a state of uniform motion in a right (that is, straight) line. Both sorts of bodies travel along

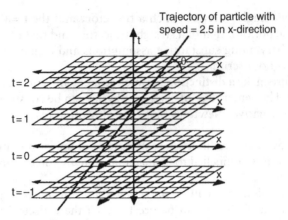

Figure 2.6   Newtonian Space-time

straight space-time trajectories. If this further distinction between bodies at absolute rest and those in uniform motion is tenable then we can use it to found the distinction between orthogonal and non-orthogonal trajectories. Bodies at rest would occupy the unique straight trajectories which are orthogonal to the simultaneity slices. The space-time which both supports the distinction between straight and curved trajectories and which allows one to define an angle between trajectories and the simultaneity hyperplanes is *Newtonian space-time*. In this space there are not only inertial frames, but special inertial frames which are at absolute rest.

Consider one such rest frame in Newtonian space-time (figure 2.6). It is easy to see that the absolute speed of a particle such as the one depicted is measured by the angle θ, the angle its trajectory makes with the planes of simultaneity. Since speed is change in distance divided by change in time, $\Delta x/\Delta t$, the instantaneous speed is equal to cot θ at any given time.

Although classical physics needed the distinction between straight and curved trajectories in space-time, the division between absolute rest and uniform motion was notoriously superfluous. According to Newtonian physics itself, no mechanical experiment could determine the absolute velocity of a body. Laboratory experiments done in a room at absolute rest would have identical outcomes to those done in a room moving with constant velocity. One can determine experimentally whether one is moving inertially or not (given knowledge of the forces at work), but no finer discriminations are empirically obtainable.

So it appears that the Newtonian picture postulates more spatiotemporal structure than actually exists. The natural response to such a situation is to retreat from the full Newtonian picture, retaining the distinction between straight and curved trajectories but abandoning that between orthogonal and non-orthogonal paths. The result is *Galilean space-time*. Galilean

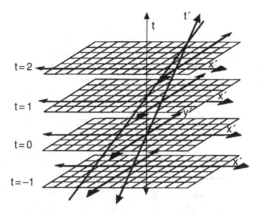

Figure 2.7   Second Coordinatization of Galilean Space-time

space-time contains just enough structure to distinguish inertial from non-inertial motion, but not enough to ground the notion of absolute velocity. In Galilean space-time Newton's first Law becomes:

> Every body continues on a straight trajectory through space-time unless some force is impressed on it.

The second Law describes how the space-time trajectory of a particle will be curved if a force is applied to the particle. Neither law need mention absolute velocities.

Figure 2.7 represents a second coordinatization of the space-time of figure 2.6. The systems of figures 2.6 and 2.7 are equally valid ways of coordinatizing a Galilean space-time, just as the two systems of figure 2.3 are equally valid Cartesian coordinates on the Euclidean plane. In the latter case, there is no "natural" orientation for the axes, and in Galilean space-time there is no "natural" time axis through the point (0, 0, 0). The time axis must be a straight trajectory, but the notion that the t axis is orthogonal while the t′ axis is tilted is purely an artifact of the figures. Nothing in the space-time singles out the t axis over t′ or over any other such axis. In drawing figure 2.7 we could just as well have "bevelled the deck" of simultaneity slices so the t′ axis would appear "vertical" as in figure 2.8. Since the angles made with the planes of simultaneity have no physical meaning, the two figures depict exactly the same situation.

Note that the particle which travels at 2.5 units per second in the unprimed coordinate system has a different velocity in the primed frame. The trajectory passes through (−1, −3, −1) and (1, −2, 1) in the primed frame. So in 2 seconds it travels 2 units in the x′ direction and 1 in the y′ for a net speed of √5/2 units per second. The t axis, which is of course at rest in the

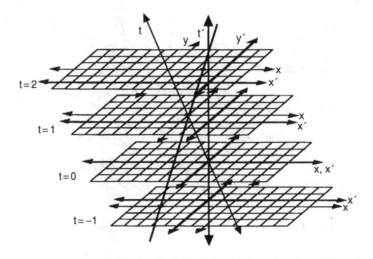

Figure 2.8   A Different Way of Drawing Figure 2.7

unprimed frame, represents the path of a particle traveling −2 units per second in the x-direction and 1 unit per second in the y-direction as judged from the t′−x′−y′ frame.

Just as there are equations for transforming from the x–y to the x′–y′ coordinates in figure 2.3, one can transform from the unprimed to the primed coordinates in Galilean space-time. The two systems shown agree on the time coordinate and on the x and y coordinate at t = 0. But as time goes on, the x and x′ axes drift ever further apart, as do the y and y′. The distance between these axes increases at a constant rate. It is easy to verify that the equations transforming between these systems are:

$$t' = t$$
$$x' = x - 2t$$
$$y' = y + t.$$

In general, the Galilean transformation between two systems (allowing for differently chosen origins) is:

$$t' = t + c_1$$
$$x' = x - v_x t + c_2$$
$$y' = y - v_y t + c_3$$

where v_x and v_y are the speeds at which the primed frame is judged to be moving from the unprimed frame.[5]

What are the invariant quantities for the Galilean transformation? The elapsed time between any two events p and q is unchanged since

$$
\begin{aligned}
\Delta t_{pq} &= t_p - t_q \\
&= (t_p + c) - (t_q + c) \\
&= (t'_p - t'_q) \\
&= \Delta t'_{pq}.
\end{aligned}
$$

A special case of the invariance of elapsed time is the agreement of all frames on whether or not two events are simultaneous, $\Delta t = 0$. Simultaneity is *absolute* in classical physics: all acceptable coordinate systems agree on which planes constitute the simultaneity slices.

Different coordinate systems in Galilean space-time also agree on the distance between *simultaneous* events, since each frame induces Cartesian coordinates on the hyperplanes of simultaneity and distance in an invariant among the Cartesean systems. But the distance between non-simultaneous events is *not* invariant. The events $(0, 0, 0)$ and $(2, 0, 0)$ in the unprimed frame, which both occur along the t axis, obviously occur in the same spatial location in the unprimed system. But in the primed frame those events are labeled $(0, 0, 0)$ and $(2, -4, 2)$: they occur at spatial locations $2\sqrt{5}$ units apart. In general, the distance between events p and q in the x-direction is $\Delta x'_{pq} = x'_p - x'_q = (x_p - v_x t_p + c) - (x_q - v_x t_q + c) = (x_p - x_q) - v_x (t_p - t_q) = \Delta x_{pq} - v_x \Delta t_{pq}$. The non-invariance of distance between non-simultaneous events is also easily seen: an object dropped in a train will be judged by passengers on the train to have traveled a rather short distance from its release to its impact on the floor, whereas according to bystanders on the ground it will have traveled some great way down track during its fall.

Since velocity in the x direction is $\Delta x / \Delta t$, it is also immediately obvious that velocity is not a Galilean invariant, as we already saw for the particle of figure 2.8. The transformation law for velocities is quite simple: a body judged to have velocity w in the x direction in the unprimed frame will have $w - v_x$ in the primed system. Similarly, a body with x-momentum mw is transformed into one with momentum $m(w - v_x)$.

If physical space has a Galilean structure, no physical law can require an object to have a specific velocity. For in the Galilean regime no body *has* a determinate absolute velocity. Any body can be given any velocity by a judicious choice of reference frames, and no frame is any more valid than any other. Ascribing a velocity to a trajectory in Galilean space-time requires the same sort of arbitrary decision as did dividing the vector from Q to P into an x-component and a y-component (figure 2.3). If only invariant quantities can represent objective features, there are no objective velocities in a Galilean world.

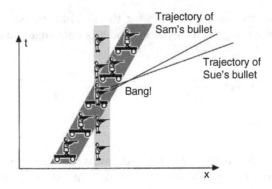

Figure 2.9   Projectiles Fired from Sources in Relative Motion

Special Relativity

In 1905 Einstein introduced a radically new theory of the structure of space and time: the Special Theory of Relativity. The Special Theory can be derived from one central postulate:

> Law of Light
> Every ray of light (in a vacuum) has the same speed, c, in all inertial frames of reference.

The fundamental feature of the Special Theory is not what it makes relative but what it makes absolute. The speed of light is an *invariant quantity* under transformations between inertial frames.

The invariance of the speed of light in a vacuum comprises two different and equally important facts. First, the speed of a light ray must be unaffected by the state of motion of its source. This is not typical behavior for any sort of projectile. For example, if Sam and Sue fire identical rifles in the x-direction, Sam from the station and Sue from a train passing by, Sue's bullet will outrun Sam's. Figure 2.9 depicts the situation from Sam's reference frame. Sue's bullet has a greater velocity than Sam's in the x-direction because it picks up an extra boost from the motion of the train.

If light behaved in this fashion, the Law of Light could not possibly hold. Since one of the bullets outruns the other they cannot possibly both be assigned some given fixed speed. If the Law of Light obtains, pulses of light emitted by Sam and Sue must, unlike the bullets, always run neck-and-neck (figure 2.10). This is important because it allows us to talk about the path that light emitted from an event *e* would take without needing to specify anything about the state of motion of the source of the light.

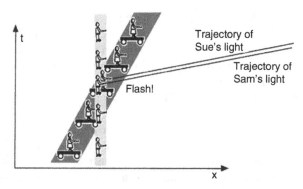

Figure 2.10 Light Emitted from Sources in Relative Motion

The independence of light trajectories from the state of light sources is a very surprising fact. It implies that there must be some objective structure which the light is tied into, a structure which determines its path. No such structure exists in Galilean space-time *per se*. One can add extra structure by introducing material bodies into the space-time; for example, the rest frame of a body of water is the unique Galilean frame in which water waves travel the same speed in all directions. But the Special Theory rejects the notion that light is a disturbance of some material medium. The structure which determines light trajectories is built into space-time itself.

Even more remarkable is the second consequence of the Law of Light. Not only must the speed of light be independent of the state of its source, it must be independent of *the inertial frame of reference in which the speed in measured*. When Sam and Sue measure the speed of a given light ray they obtain the same answer even though they are in motion relative to one another. But as we have seen, under the Galilean transformation *no* velocity is invariant. So for the Law of Light to hold, the Galilean transformation cannot represent the transition between inertial frames, and the structure of space-time implicit in that transformation must be rejected.

Consider how one might calculate the speed of a light ray. Suppose Sue stands at one end of her train car and emits a pulse of light (event q), which travels across the car and some short time later hits the far wall (event p). Since it travels in the x-direction, its speed will be $\Delta x_{pq}/\Delta t_{pq}$. Under a Galilean transformation Δx will be changed while Δt remains invariant, so the speed must also change. For $\Delta x_{pq}/\Delta t_{pq}$ to remain *unchanged* under a coordinate transformation either Δt and Δx must both be unchanged or they must both change in the same proportion. For the Law of Light to hold, this proportionality need not obtain for all pairs of events p and q but only for pairs of events which are parts of the trajectory of a light ray.

It is obvious that Δt and Δx cannot both be invariants. When Sue performs her experiment she judges that the light has traveled only the distance across her train car, while Sam will see it as traveling much further as the car travels down the tracks. So we get our first qualitative result concerning coordinate transformations in Special Relativity: in transformations between relatively moving inertial frames the elapsed time between events is *not* an invariant quantity.

How, then, do Sam's coordinates relate to Sue's? One can derive the relevant transformation more or less rigorously; I will provide a somewhat loose and sketchy argument. Let's return to a full four-dimensional space-time. Suppose the point p has coordinates (0, 0, 0, 0) in both Sam's (unprimed) and Sue's (primed) coordinate systems. The point q, where the light is absorbed, will have coordinates (t, x, 0, 0) and (t', x', 0, 0) respectively. For the speed of light to be constant, $\Delta x/\Delta t = x/t = \Delta x'/\Delta t' = x'/t' = c$. We begin with the Galilean transformation $x' = x - vt$, where v is the relative velocity of the two reference frames (we suppose they move relatively to one another in the x-direction). The new transformation must reduce to the Galilean transformation when the relative velocity is low since Newtonian physics is very nearly correct in that regime. So let the new transformation have the form $x' = \gamma(x - vt)$, with γ being some correction factor which depends on the relative velocity v. γ must approach 1 as v tends to 0. The corresponding transformation from x' back to x is $x = \gamma(x' + vt')$ (Sam sees Sue as traveling in the positive x-direction, Sue sees Sam as traveling equally fast in the negative x-direction).

Substituting x = ct and x' = ct' from above, these equations become $ct' = \gamma(ct - vt) = \gamma t(c - v)$ and $ct = \gamma(ct' + vt') = \gamma t'(c + v)$. Multiplying these equations together yields $c^2 tt' = \gamma^2 tt'(c^2 - v^2)$ or $c^2 = \gamma^2(c^2 - v^2)$. Hence $\gamma^2 = c^2/c^2 - v^2 = 1/(1 - v^2/c^2)$. The transformation for spatial coordinate x is therefore:

$$x' = \frac{1}{\sqrt{1 - \dfrac{v^2}{c^2}}}(x - vt)$$

As desired, when $v \to 0$, $v^2/c^2 \to 0$, and $\gamma \to 1$, yielding the Galilean transformation. But as v approaches the speed of light $v^2/c^2 \to 1$ and γ tends toward infinity. The spatial distance between two events becomes larger under this transformation than it would be under a Galilean transformation.

What about time? Taking even more liberties, we can derive the time transformation as follows. Starting with $x' = ct' = \gamma(x - vt)$ we substitute x = ct and t = x/c to get $ct' = \gamma(ct - vx/c) = \gamma c(t - vx/c^2)$ so

$t' = \gamma(t - vx/c^2)$.

Again as $v \to 0$, $vx/c^2 \to 0$, and $\gamma \to 1$, returning the Galilean transformation $t' = t$.

Putting these results together (given that motion in the x-direction cannot affect the y or z coordinates) we get the *Lorentz transformation*:

$$t' = \gamma(t - vx/c^2)$$
$$x' = \gamma(x - vt) \quad \left(\gamma = (1 - v^2/c^2)^{-\frac{1}{2}} \right)$$
$$y' = y$$
$$z' = z.$$

These transformations hold when the two coordinate systems share an origin and the primed system moves with speed v in the positive *x*-direction relative to the unprimed system.

As expected, the amount of time elapsed between p and q is not invariant under the Lorentz transformation. This explains how the speed of light can remain constant. But the disagreement between the primed and unprimed coordinates goes far beyond a dispute over time *scales*. Most of our interest in the Special Theory derives from the remarkable implications of the relation $t' = \gamma(t - vx/c^2)$. We will therefore devote some care to exploring the meaning of this transformation.

To simplify matters for the moment, consider a two-dimensional space-time coordinatized by x and t. We choose units such that c = 1: a light ray travels one space unit per time unit. Light will therefore follow trajectories tilted at 45° with respect to the x and t axes if we draw the space-time in the usual way. A flashbulb set off at the origin would send out two flashes of light, one traveling to the right and the other to the left. These trajectories form the "light cone" at the origin. In a two-dimensional space-time the light cone is reduced to a pair of straight lines. In a three-dimensional space-time, with two spatial dimensions, the lightbulb would produce an expanding circle of light whose trajectory forms a cone in the space-time. A real lightbulb produces an expanding sphere of light which we cannot represent using a pictorial space-time diagram. The light cone can be extrapolated back into times before the event in question, where it represents the path of light which will arrive at, rather than originate from, the event (see figure 2.11). The x axis represents all of the points with coordinate t = 0, all of the events that are simultaneous with event P. In general, the x–t system carves up the space-time into simultaneity slices as depicted in figure 2.12.

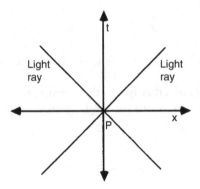

Figure 2.11   The Light Cone in the x–t Frame

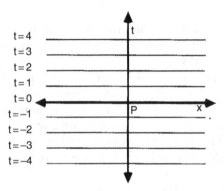

Figure 2.12   Simultaneity Slices in the x–t Frame

But in the x′–t′ frame things look very different. The events judged to be simultaneous with P in *that* frame are those with the coordinate t′=0. Since $t' = \gamma(t - vx/c^2)$, the set of events simultaneous with P in this frame are those with coordinates $t = vx/c^2$ in the x–t frame. This set of points forms a straight line through P with slope v/c^2, as depicted in figure 2.13.

We can now see that *simultaneity is relative to a reference frame*. According to the x–t frame the events marked P and Q take place at the same time and R occurs later. But according to the x′–t′ frame the events marked P and R take place at the same time, a time at which Q has already occurred! The x′–t′ frame divides up the space-time into a different set of simultaneity slices, those depicted in figure 2.14.

Finally, we can draw in the t′ axis, which is just the trajectory of the spatial origin of the x′–t′ frame. Since that frame is moving to the right, the t′ axis is a line tilted to the right. Transforming from the x–t to the x′–t′

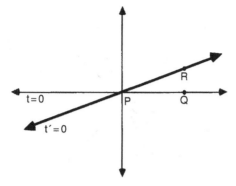

Figure 2.13 Points Simultaneous with P in the x′–t′ Frame

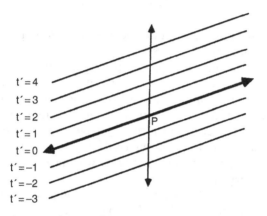

Figure 2.14 Simultaneity Slices in the x′–t′ Frame

axes is rather like rotating the axes in regular Euclidean space, except that as the x′ axis comes up the t′ axis tilts down (figure 2.15). This actually is what we should expect: if light is to have the same velocity, 1, in all coordinate systems then the light ray should "split the difference" between the x and t axes in every frame. This is how the Lorentz transformation keeps the speed of light constant.

Consequences of the Lorentz Transformation

All of the surprising and counterintuitive aspects of Special Relativity flow from the Lorentz transformation. To appreciate these consequences one must bear two things in mind. First, the Lorentz transformation shows that

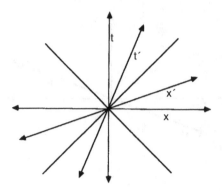

Figure 2.15   The x–t and x′–t′ Axes

how the spatio-temporal separations between events are parsed into spatial and temporal components differs greatly in different frames of reference. Second, our fundamental interpretive principle is that none of the reference frames is preferred by nature over any of the others. On the one hand, this provides us with many, oddly different, accounts of the spatial and temporal structure of the world. On the other, it vastly reduces the number of objective, frame-independent facts. We have already remarked that the frame-independent facts are to be rendered by *invariant* quantities and laws. Before turning to these invariants, let's review some of the quantities which are no longer invariant.

We have already seen that simultaneity is no longer absolute. The relativity of the time order of events will play an essential role in all of our results, so its limits should be noted. Returning from a two-dimensional to a three-dimensional space-time, the light cone at each point is represented as a cone, as in figure 2.16. The light cone at each point divides the space-time into three regions. Points in the future-directed light cone of an event such as P are in the *absolute future* of that point. The *time order* between P and any event in its future light cone is invariant: all Lorentz frames will agree that those events come after P in time. Similarly, all reference frames will agree that the events in the *absolute past* of P happened prior to P in time.[6] But the time order between P and any event outside its light cone, such as Q, is not Lorentz invariant. In the t–x–y frame, P and Q are judged to be simultaneous, while in the t′–x′–y′ frame P comes before Q and in the t″–x″–y″ frame P comes after Q. So there is no objective fact of the matter about which, if either, of the two events occurred first.

The spatial dimensions of rigid bodies are also altered by the Lorentz transformation. An object will appear longer in a frame of reference in which it is at rest than in a frame in which it is moving. This *Lorentz-Fitzgerald*

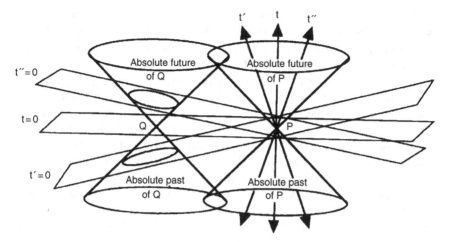

Figure 2.16   Light Cones of Two Events

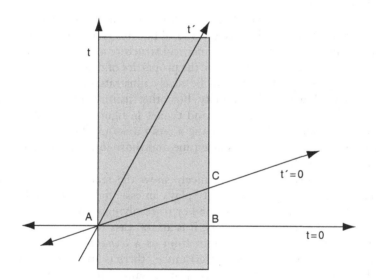

Figure 2.17   Rigid Bar and Two Reference Frames

contraction may at first appear to be the opposite of what the transformation equations predict. Consider the bar depicted in figure 2.17. It is at rest in the t–x frame, and its length can be determined in that frame by calculating the spatial separation between an event on its right end and one on its left end, such as A and B. Suppose that A is the origin of the frame and that the bar has length d in that frame. Then the t–x coordinates of A and B are (0, 0) and (0, d) respectively. When we transform to the primed frame, according to which the bar is moving, the spatial distance between A and B *increases* by a factor of $(1 - v^2/c^2)^{-\frac{1}{2}}$, as we find by plugging the coordinates into the Lorentz formula. Why then does the bar contract?

The key lies in observing that the length of the bar in a frame in which it is moving is the spatial separation between *simultaneous* events on its two edges (as judged in that frame). In the primed frame A and B are no longer judged to be simultaneous, and we must locate the event C which does occur at the same time as A in this frame. If A is also the origin of the primed frame, this amounts to finding the point along the right edge of the bar for which $t' = 0$. All of the points along the right edge have x-coordinate d. Using the transformation $t' = \gamma(t - vx/c^2)$ we now seek the *unprimed* coordinates of the point C. We can easily solve $0 = \gamma(t - vd/c^2)$ to get $t = vd/c^2$, so the t – x coordinates of C are $(vd/c^2, d)$. Using $x' = \gamma(x - vt)$, we see that the coordinates of C in the primed frame are $(0, \gamma(d - dv^2/c^2)) = (0, d\sqrt{(1 - v^2/c^2)})$, and the length of the bar *shrinks* by $\sqrt{(1 - v^2/c^2)}$ or $1/\gamma$. As $v \to c$, the length of the bar approaches 0.

It may seem wrong, given figure 2.17, to say that the distance between A and C is less than that between A and B, but this is because the paper on which the figure is printed has the metrical structure of the Euclidean plane and therefore can not directly mimic the properties of relativistic space-time. The relativity of simultaneity can be nicely illustrated by drawing in the various hyperplanes of simultaneity, but other features of the picture can be deceiving. For example, the x' and t' axes in figures 2.15 and 2.17 are orthogonal to one another, just as the x and t axes are. Because of the differences between rotations in space-time and those in Euclidean space this cannot be captured.

The illustrations do, as noted, nicely show the relativity of simultaneity, and the relativity of simultaneity plays an essential role in the Lorentz-Fitzgerald contraction. It is also the key to resolving some of the apparent paradoxes of the relativistic effects. It is all too easy to fall into the trap of thinking of relativistic length contraction as a *dynamical effect* of *going fast*: the faster you go, the more "resistance" there is and the more you get squashed. But sublight velocities have just the same status in relativistic as in Galilean space-time: there is no fact of the matter about how fast anything is going, it all depends on the reference frame. So to the person in the station the train is judged as moving, to the person on the train the station is judged as moving, and neither is more correct.

This leads to a seeming contradiction. Take a car and a tunnel which, when at rest relative to one another, are exactly the same length. Now get in the car and drive it through the tunnel. According to the tunnel, the car is moving and therefore suffers a contraction: the car should fit entirely inside

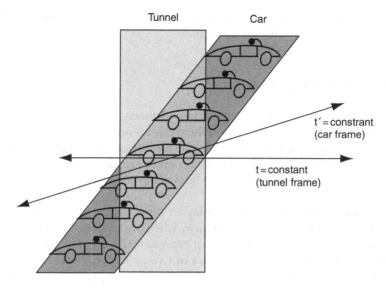

Figure 2.18   Car and Tunnel with Two Reference Frames

the tunnel. On the other hand, according to the car the tunnel is moving, so the car should now be longer than the tunnel. No matter how objects shrink or grow, how can it both be the case that the car is longer than the tunnel and that it is shorter than the tunnel?

Figure 2.18 depicts the situation, shown in the frame of reference of the tunnel. In that reference frame there is a moment, a simultaneity slice t = constant, at which the front end and rear end of the car are both inside the tunnel, so the car must (in that reference frame) be shorter than the tunnel. But from the car's point of view (or the point of view of the inertial frame in which the car is at rest) there is an instant t′ = constant at which the front end of the car has already emerged from the tunnel while the rear end has not yet entered, hence according to the driver the car must be longer than the tunnel.

The story of the car and the tunnel illustrates how seemingly non-temporal notions may be infected by hidden temporal aspects. The question of whether the car or the tunnel is longer appears to be a question purely about the spatial qualities of the two objects. If it were, then the differing evaluations of the two reference frames would constitute a true contradiction: the car and the tunnel cannot each be longer than the other. But once we see that the car being longer than the tunnel is a matter of the front and back ends of the car being outside the tunnel *at the same instant*, it becomes clear how the relativity of simultaneity resolves the tension between the two judgments. Since the different observers disagree on which sets of events constitute an instant, they may disagree on whether there is an instant when every part of the car is within the tunnel.

Lorentz Invariant Quantities

Under the Lorentz transformation spatial distance between events does not remain invariant. The time elapsed between events is not invariant. Time order is not invariant for events which lie outside one another's light cones. The velocity of light is invariant, but no other velocity is. So if we look to invariant quantities to find objective facts about the world we seem to have precious little to work with.

The case is, however, quite similar to regular rotation of coordinate frames in Euclidean space. As we saw, when we rotate our spatial axes, the quantities Δx_{PQ}, and Δy_{PQ}, which represent the decomposition of the vector between P and Q into the sum of two orthogonal vectors, change. But the distance $\sqrt{\Delta x^2 + \Delta y^2}$ is invariant. It is this reality, the distance, which various coordinate systems resolve differently into components Δx and Δy.

Similarly, different inertial frames break up space-time differently into spatial and temporal components, but there is still an invariant "distance" between events. Take any two events P and Q, and consider the quantity $c^2 \Delta t^2 - \Delta x^2 - \Delta y^2 - \Delta z^2$ (in some reference system). Under the Lorentz transformation

$$\Delta t'^2 = \gamma^2 (\Delta t - v\Delta x / c^2)^2, \quad \Delta x'^2 = \gamma^2 (\Delta x - v\Delta t)^2,$$
$$\Delta y'^2 = \Delta y^2, \quad \Delta z'^2 = \Delta z^2,$$

so

$$
\begin{aligned}
& c^2 \Delta t'^2 - \Delta x'^2 - \Delta y'^2 - \Delta z'^2 \\
& = c^2 \gamma^2 (\Delta t^2 - 2v\Delta t\Delta x / c^2 + v^2 \Delta x^2 / c^4) + \gamma^2 (\Delta x^2 - 2v\Delta t\Delta x + v^2 \Delta t^2) \\
& \quad - \Delta y^2 - \Delta z^2 \\
& = \gamma^2 (c^2 \Delta t^2 - 2v\Delta t\Delta x + v^2 \Delta x^2 / c^2 + \Delta x^2 - 2v\Delta t\Delta x + v^2 \Delta t^2) - \Delta y^2 - \Delta z^2 \\
& = \gamma^2 (c^2 \Delta t^2 (1 - v^2 / c^2) - \Delta x^2 (1 - v^2 / c^2)) - \Delta y^2 - \Delta z^2 \\
& = \gamma^2 (1 - v^2 / c^2)(c^2 \Delta t^2 - \Delta x^2) - \Delta y^2 - \Delta z^2 \\
& = c^2 \Delta t^2 - \Delta x^2 - \Delta y^2 - \Delta z^2.
\end{aligned}
$$

This invariant quantity is the underlying reality, the true spatio-temporal structure which is only arbitrarily divided up into space and time by coordinate frames. We may lay down some set of coordinates t, x, y, and z, but, as Minkowski put it:

we may also designate time t′, but then must of necessity, in connexion therewith, define space by the parameters x′, y′, z′, in which case physical laws would be expressed in exactly the same way by means of x′, y′, z′, t′, as by means of x, y, z, t. We should then have in the world not *space*, but an infinite number of spaces, analogously as there are in three-dimensional space an infinite number of planes. Three-dimensional geometry becomes a chapter in four-dimensional physics. (Minkowski 1952, pp. 79–80)

Henceforth space by itself, and time by itself, are doomed to fade away into mere shadows, and only a kind of union of the two will preserve an independent reality. (ibid., p. 75)

The analogy between the invariant interval $c^2 \Delta t^2 - \Delta x^2 - \Delta y^2 - \Delta z^2$ and the Euclidean distance $\sqrt{(\Delta x^2 + \Delta y^2)}$ is not perfect. In Euclidean space no two distinct points can have zero distance between them. But the invariant interval between any event and any other event on its light cone is zero. For example, the velocity of a light ray propagating in the x-direction is $c = \Delta x / \Delta t$, yielding the equation $\Delta x = c\Delta t$. Δy and Δz are both zero. Plugging this into the definition of the invariant interval gives $c^2 \Delta t^2 - \Delta x^2 - \Delta y^2 - \Delta z^2 = c^2 \Delta t^2 - c^2 \Delta t^2 - 0 - 0 = 0$. Indeed, the light cone of an event P may be defined as the set of all points Q such that $c^2 \Delta t_{PQ}{}^2 - \Delta x_{PQ}{}^2 - \Delta y_{PQ}{}^2 - \Delta z_{PQ}{}^2$ is zero. It is then obvious that all coordinate frames agree on the light cone structure of space-time since they agree on the invariant interval between events.

Invariant interval zero cannot, then, be assimilated to Euclidean distance zero. Things get even more difficult for events P and Q that lie outside one another's light cones. In that case $c^2 \Delta t_{PQ}{}^2 - \Delta x_{PQ}{}^2 - \Delta y_{PQ}{}^2 - \Delta z_{PQ}{}^2$ is less than zero, and no comparison with distances is possible. Pairs of points whose interval is less than zero are said to be *space-like* separated. If the interval is zero they are *light-like* separated, and if it is positive *time-like* separated. For any pair of space-like separated points there exists an inertial frame in which they are simultaneous, and so only differ in space coordinates, and for any pair of time-like separated points there is a frame in which they have identical space coordinates and differ only in time.

The connection between time-like separation and time deserves some comment. A *time-like trajectory* is one which "threads the light cones," never inclining as much as 45° in our diagram. A particle traversing a time-like trajectory never reaches the speed of light: the invariant interval between any two points on a time-like trajectory is always positive. We can therefore always extract the square root of the interval $\sqrt{(c^2 \Delta t^2 - \Delta x^2 - \Delta y^2 - \Delta z^2)}$ and have a real positive quantity. Integrating this quantity along a time-like trajectory can give us a measure of the "length" of a time-like path.

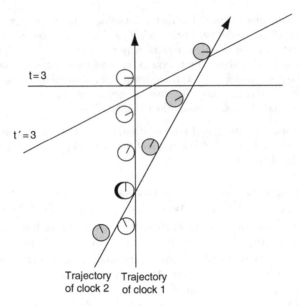

Figure 2.19   Symmetry of Time Dilation

The length of a path so defined is not merely a concocted invariant quantity. A fundamental postulate in the interpretation of Relativity is that:

A clock measures the quantity

$$\frac{1}{c} \sqrt{c^2 \, \Delta t^2 - \Delta x^2 - \Delta y^2 - \Delta z^2}$$

along the trajectory it travels.

The ticks of a clock mark off portions of equal "length" along that trajectory. For example, in its own inertial frame a clock is at rest, so Δx, Δy and Δz are all zero. The invariant length of the trajectory of the clock is therefore $\sqrt{(c^2 \Delta t^2)} = c \Delta t$. The clock evidently measures this quantity divided by c.

We can now understand one of the most famous relativistic effects: time dilation. A clock is judged to run slower in a frame of reference in which it is moving than the clocks at rest in that frame. As with the Lorentz-Fitzgerald contraction, the most puzzling aspect of the phenomenon is its symmetry: each of the two relatively moving clocks (in its frame) judges the other to be running slow. And again the solution lies in the relativity of simultaneity. As figure 2.19 shows, in the $t-x$ frame, when clock 1 has ticked off three seconds, clock 2 has not yet reached its third tick, and in the $t'-x'$ frame when clock 2 has reached it third tick, clock 1 has not yet reached that mark.

This completes our introduction to the Special Theory of Relativity. For most of our purposes, the graphical representation of space-time will suffice, but one caveat must be heeded. Whenever we draw a diagram we arbitrarily select one time-like direction to be vertical and its corresponding simultaneity hypersurfaces to be horizontal. This choice does not reflect any objective intrinsic difference between that time-like direction and any other. This is the content of the relativity principle.

A test of how well one has internalized the spirit of Relativity is how one is tempted to answer the question: what would it be like to travel at 99.99 percent the speed of light?

A bad answer: due to the Lorentz-Fitzgerald contraction and to time dilation, one would feel squashed flat as a pancake and see time passing very slowly.

A marginally better answer: one would indeed be squashed flat as a pancake and one's clocks would be running slowly, but this would not be noticeable. For one's meter sticks would have shrunk proportionally, and so would measure things as being their original size. And one's brain processes would have correspondingly slowed, so the retardation of the clocks would not be noticeable.

The truly relativistic answer: right now you *are* traveling at 99.99 percent the speed of light – in some perfectly legitimate inertial frame. It is no more correct to say that you are now at rest than that you are infinitesimally close to moving at light speed. Indeed, except for things traveling at light speed, it makes no sense to attribute any absolute velocity to any object.[7] Not being invariant, sublight velocities are not real properties of objects. There can therefore be no dynamical effect of "traveling near the speed of light" because there is no such objective state as traveling near the speed of light. Things don't shrink or slow down. Rather, there are always infinitely many ways of expressing space-time intervals in terms of distance in space and elapsed time. None of these ways, represented by the various inertial frames, is any more valid, or less valid, than any other.

Notes

1 Since the experimental verification of violations of Bell's inequality, a cottage industry has sprung up among philosophers of physics which is dedicated to distinguishing various senses of "locality." Michael Redhead lists seven different sorts (1987, pp. 77, 82, 113, 139, 141), Arthur Fine a more modest two (1986, pp. 59–63), Geoff Hellman three (1982, pp. 465–6), and so on. The task of correlating these various senses of locality, while worthwhile, would not much further our present project. Many of these distinctions will be recognized, if implicitly, in our discussion. Others reflect interests which fall outside of our own. Some of Redhead's different construals of locality, for example, turn on

whether the distant particle is affected by (1) changing an unsharp value for a quantity into a sharp value; (2) changing an undefined value into a defined value; or (3) changing a sharp value into another sharp value. All of these involve some sort of non-local influence, and we will not be particularly concerned with which mechanism is at work.

2   The continuity requirement is not always respected in some familiar coordinate systems. In polar coordinates, for example, there is a coordinate discontinuity in the polar angle along $\theta = 0°$. Some manifolds, such as the sphere, cannot be covered by a single system which has no coordinate singularities. In this case, one covers the manifold with a set of partial coordinate systems called *charts*, each of which satisfies the continuity requirement.

3   This is not strictly true. In the example given exactly one point, the *fixed point* of the transformation from one system to the other, is assigned the same coordinates in both systems. The fixed point for this pair is near (5.232, −4.598).

4   All of the classical space-times we will discuss, and more, are presented and discussed in John Earman's definitive study *World Enough and Space-Time* (1989). Our presentation will be less technical than Earman's; those interested in the details should consult his text. Nice presentations of Galilean space-time (although not under that name) and of the General Theory of Relativity can also be found in Geroch (1978).

5   The full Galilean group includes a wider set of transformations than those mentioned here, as the spatial axes of the two systems could be differently oriented. The invariants of the full Galilean group are physically quite important, but are aside from our present concerns.

6   Although time *order* is invariant in these cases, the total time elapsed between the events is not, as is clear from the transformation.

7   Although this sounds exactly like the situation in Galilean space-time, it is important to note the differences. In Galilean space-time *there is no such thing as the angle between a hypersurface of simultaneity and a trajectory*. Thus absolute velocities cannot be defined as in Newtonian space-time. But in Minkowski space-time, there always is a definite invariant angle between a given hyperplane and a trajectory. The failure of absolute velocities to exist lies not here but rather in the existence of a multiplicity of hyperplanes of simultaneity. In many respects, Minkowski space-time is more similar to Newtonian than to Galilean space-time: both Newtonian and relativistic space-time are metric spaces, with "distances" defined between all pairs of points and angles between all pairs of vectors. Galilean space-time has a weaker structure.

3

Finger Exercise: Superluminal Matter Transport

The State of Play

We are now in a position to understand the significance of the design of Aspect's experiment testing the violation of Bell's inequality. In the frame of reference of the laboratory, the two wings of the experiment were about 12 meters apart and the polarization direction being measured was changed every 10^{-8} seconds. In the frame of reference of the laboratory the two particles arrive at their respective detectors essentially simultaneously. From the latter condition it follows that the detection events were at space-like separation from one another. From the former, since the speed of light is only 3×10^8 meters per second, it follows that the *setting* of the detector for the particle on one wing was space-like separated from the detection of the corresponding particle on the other. So even a light signal sent at the moment the right detector was set (before any particle arrived) would only arrive at the left detector after the left-hand particle was observed. Such a signal would be too late to have any effect on the result on the left, as can be seen in figure 3.1 Similarly, a light ray sent from the left at the moment of setting could not arrive in time to affect the observation on the right (consider the mirror-image space-time diagram).

Our fundamental puzzle is now complete. If no sort of influence can travel faster than light, or can directly connect space-like separated events, then there is no way for the observation of one photon to have any effect on the other. If you and your partner manage to secrete radio transmitters on your persons in order to communicate which question has been asked of you, it will be of no avail since each of you will be required to answer your own question before

Quantum Non-Locality and Relativity: Metaphysical Intimations of Modern Physics, Third Edition. Tim Maudlin.
© 2011 Tim Maudlin. Published 2011 by Blackwell Publishing Ltd.

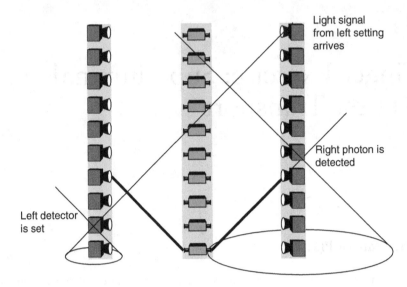

Figure 3.1   Aspect's Experiment in Space-time

your partner's report has arrived. Furthermore, even if you manage to station a confederate at each detector whose job is to radio the setting of the detector as soon as it is made, that will be of no help for the same reason.

We have already entered a realm of unrealistic speculation since the photons cannot have helpful spies already in position in the laboratory, but just to close off more loopholes, let's push the question a bit further. First, we should note that the idea that the mere event of *setting* one detector could have any direct effect on the distant photon is quite farfetched. We saw that the quantum connection is discriminating: if we send out a hundred pairs of particles at once, with a hundred detectors on each wing, the result for each particle will depend only on the observation carried out on its twin. According to quantum mechanics, the other 99 observations on the distant wing, the other 99 settings of detectors, will have no effect at all. So no general influence of detector settings on distant particles can be postulated: only the setting of the detector *which the twin will hit* is relevant. But before it is hit, there is nothing that especially distinguishes the relevant detector from the others, so it is hard to imagine that setting *it* would have any special influence on one particular distant photon. But let us grant a pre-existing affinity of a particular detector for a particular distant photon and press on.

On the assumption that no influence travels faster than light, in order to arrive in time the message saying what the setting will be must be sent even before that setting has been made. How could this be done? One might be able to figure out what the setting will be if one knows enough about how the choice will be made. Suppose, for example, that a computer will choose the setting, based on the solution to some algebraic equation. Sufficient information about the construction and program of the distant computer might lie in the back light cone of the detection event for the setting of the distant detector to be inferred. If so, then radio signals could convey the requisite information (since they could convey information about anything in the back light cone). Now one problem which arises is that the setting of the distant detector could be determined by any number of means. It could be coupled to a computer, or a throw of dice, or the stock market, or the whim of the experimenter, and so on. So if the behavior of each photon is influenced only by events in its past light cone, there would have to be fundamental laws connecting the photon's trajectory to distant facts about the disposition of dice and the number of shares of IBM being sold and so on, a different connection for each possible mechanism for determining the setting. This strains credulity.

Bell himself addressed this question, arguing that the setting of the detectors can reasonably be treated as free variables (Bell 1987, ch. 12). The reason is not because the detectors must be set by truly stochastic means: most "random number" generators are perfectly deterministic systems. But the facts which determine the output of such generators are so scattered and miscellaneous, just as the facts at 12:00 which determine how many shares of IBM will be traded at 12:05 are scattered and miscellaneous, that it becomes impossible to imagine that exactly those conditions would have some *other* notable effect (such as directly influencing distant photons).

A final nail in the coffin of such conspiracy theories come from the observation that one could arrange things so that no amount of information in the back light cone of the detection event, no matter how scattered and obscure, could suffice to determine the setting of the distant apparatus. To accomplish this we need to have the setting of each polarizer determined by incoming electromagnetic radiation left over by the Big Bang. If the radiation observatory at each wing is pointed away from the other wing, then the entire past light cone of each measurement will not contain enough information to determine the distant setting. Of course, we have already got beyond any experiment that would ever have funding. But in this case there is simply no means by which subluminal or luminal signals could provide information about the setting on one side of the experiment to the other side. Even if we allow each photon to base its response on the totality of events in its past light cone, no strategy can be devised which will reliably return the quantum correlations. For the vital piece of information, the setting of the distant detector, is not contained in that back light cone. Yet quantum mechanics predicts that the observed correlations would remain even in this setting.

According to the terminology we have adopted, a *local* theory is a theory according to which no causal influence[1] can propagate faster than light, where no event can have an effect of any sort outside its future light cone. Such theories could be either deterministic or stochastic. If stochastic, then the total state of the world in the back light cone of a region of space-time can, along with the laws, fix only the probability of events of various sorts occurring in that region. But if the theory is local, no information about space-like separated events could improve upon that prediction. Such events could have no direct influence, and by conditionalizing on the entire back light cone we would have screened off any space-like correlations which derive from common causes, since common causes would have to reside in the back light cone. (Two correlated stochastic events are said to be screened off from one another by a third event if they become probabilistically independent once one conditionalizes on the third event). As we have seen, no local stochastic theory could get the perfect correlation of our photons (when the detectors have the same polarization) right. If the detection event on the right involves some irreducibly random process, the outcome of that random process could not have any influence on the left, so the left-hand photon could not be nomically guaranteed to register the same result. And no local deterministic theory can get the rest of the quantum correlations right since the setting of the distant apparatus (or any other facts which determine that setting) need not be in the past light cone of the observation event.

Spatial separation would be a perfect insulator from causal influence if no influence could go faster than light. Since Aspect's experiment shows no such insulating effects, some direct causal connections must exist between space-like separated events. But we might now wonder why this result should provoke any anxiety about conflicts with Relativity. For in deriving the Lorentz transformation and explicating the relativistic space-time structure we nowhere assumed that the speed of light acts as any sort of a limit on anything. We used only the *invariance* of the speed of light as encapsulated in the Law of Light. Nonetheless, Relativity is commonly taken to imply some sort of restriction on how fast something can go. And given the fundamental difference between space-like and time-like separated events, one might suspect that connections between space-like separated events may display rather peculiar and counterintuitive properties. Our job now, and in the following three chapters, is to explore what sort of limit the speed of light might be. On the quantum non-locality side, we also want to determine what sort of limit the speed of light cannot be.

Particles and Relativistic Mass Increase

In the preceding chapter we derived the Lorentz transformation and noted various relativistic effects which concern the description of the world in terms of space and time. We did not touch on one of the more famous relativistic

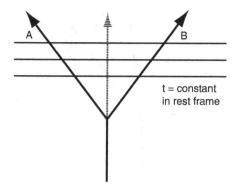

Figure 3.2 The Fission in Space-time

effects: the relativistic mass increase. As it is usually put, the faster particles go, the heavier they get. This formulation is subject to the same caveats as the analogous ones concerning the time-dilation and length-contraction effects: there is no absolute fact of the matter about how fast any subluminal particle is going, nor about which of two particles is going faster, and so no corresponding fact about how massive a particle is or which of the two has more mass. All such judgments depend upon the reference frame.

So far, we only have results about the frame-dependence of spatio-temporal quantities. In order to tie mass to space-time structure, one needs a dynamical principle involving mass which is to hold in all reference frames. For our example we will use the principle that the center of mass of a system which is not subject to external forces should, in all inertial frames, travel inertially. This is just an instance of Newton's First Law, applied to a compound system subject to no external forces. In such a system, the center of mass travels in a straight line through space-time, where the center of mass in a given dimension is calculated by multiplying the position of each particle by its mass, adding up these products, and dividing by the total mass. For example, the center of mass of a system consisting of a 4 pound object located 2 feet to the left of the origin in the x-direction and a 3 pound object located 3 feet to the right of the origin in the x-direction is $(-8 + 9)/7 = 1/7$ of a foot to the right of the origin.

Consider, in a two-dimensional space-time, an object which fissions symmetrically in its rest frame, e.g. a bomb splitting into two pieces which travel off with equal speeds in the positive and negative x-directions, as judged from the rest frame. Call the two pieces A and B. The space-time diagram of the fission would look like figure 3.2.

The first thing we prove is that in the rest frame M_A, the mass of the left-moving piece, equals M_B, the mass of the right-moving one. This is obvious by symmetry, but it will be useful to prove it in order to illustrate the general line

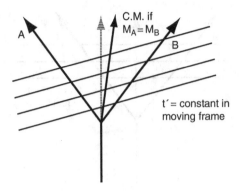

Figure 3.3   Fission with Simultaneity Slices of Moving Frame

of reasoning. Since the center of mass of the system moves inertially, it must continue along the dotted trajectory. At any given moment after the fission, the whole system consists of A and B which are, in the rest frame, judged to be equal distances from the dotted trajectory. So for the center of mass to lie on that trajectory, the two masses must, in the rest frame, be equal.

What happens when we switch to a frame of reference moving to the right? The simultaneity slices tip up to the right, as in figure 3.3. If the masses of the two pieces remain equal in the new frame of reference, then one would calculate the center of mass in just the same way as in the rest frame, by bisecting, at each instant, the line connecting the world-line of A with the worldline of B. But bisecting these tilted lines yields the incorrect trajectory of the center of mass indicated by the arrow in figure 3.3. In order for the center of mass to remain on the inertial trajectory in the moving frame, M_A must be greater than M_B in that frame. If A is sufficiently more massive, that will pull the center of mass back to its proper path. Since A is moving faster than B in a frame of reference moving to the right, we have now got the qualitative result: given two objects with the same rest mass, that which moves faster in some frame of reference will be judged to be more massive in that frame. (The equality of the rest masses of A and B is established by the symmetry of the problem.) Once again, the relativity of simultaneity lies at the heart of the problem: because of the different orientations of simultaneity slices in different frames, masses of objects must change from frame to frame to keep the Law of Inertia intact.

Furthermore, by simple geometrical considerations we can calculate the ratio of the masses of A and B in any given frame, and derive the exact Relativistic formula. We will do the calculation using only the x–t coordinates of the rest frame. Consider some frame of reference whose simultaneity slices have slope s in the rest-frame coordinates. Choose the simultaneity slice of that frame which passes through the point P = (–V, 1), i.e. the point on the worldline of A which occurs one second after the fission, as judged

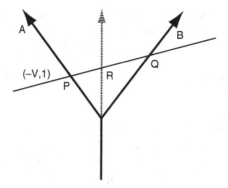

Figure 3.4 Calculating the Ratio of Masses

from the rest frame (see figure 3.4). We wish to find the coordinates of the point Q where the simultaneity slice intersects the worldline of B. Since the simultaneity slice has slope s, its equation is $t = sx + (1+sV)$, where V is the speed of A and B in the rest frame. The equation for the worldline of B is $x = Vt$. So at the point of intersection, $t = sVt + (1 + sV)$, or $t(1 - sV) = (1 + sV)$, or $t = \dfrac{1+sV}{1-sV}$. The simultaneity slice therefore intersects B at $\left(V\dfrac{1+sV}{1-sV}, \dfrac{1+sV}{1-sV} \right)$. The coordinates of the point R are $(0, 1 + sV)$.

Evidently, if the space-time, were a Euclidean space the ratio of PR to RQ would be $1 : \dfrac{1+sV}{1-sV}$. And even in Minkowski space-time, since PR and RQ are collinear, in every frame the spatial and temporal distances between P and R and between R and Q stand in the same ratio, as can be proved by direct calculation. The Relativistic interval is $(\Delta t^2 - \Delta x^2)$ in any frame of reference. The interval between P and R is therefore

$$(sV)^2 - (V)^2.$$

The interval between R and Q is

$$\left(\frac{1+sV}{1-sV} - (1+sV) \right)^2 - \left(V\frac{1+sV}{1-sV} \right)^2 =$$

$$\left(\frac{1+sV - 1 + (sV)^2}{1-sV} \right)^2 - \left(V\frac{1+sV}{1-sV} \right)^2 =$$

$$\left(\frac{1+sV}{1-sV}(sV) \right)^2 - \left(\frac{1+sV}{1-sV}V \right)^2 =$$

$$\left(\frac{1+sV}{1-sV} \right)^2 ((sV)^2 - (V)^2).$$

In the moving frame, P, R, and Q are simultaneous, so the Relativistic interval between them is just the negative of the square of their spatial separation, which remains in the ratio $1 : \dfrac{1+sV}{1-sV}$. So for the center of mass in the moving frame to remain on the inertial trajectory, the mass ratio of A to B must be $\dfrac{1+sV}{1-sV} : 1$.

To get the precise Relativistic formula, we now consider the rest frame of particle B. Since the slope of the worldline of B is $1/V$, the slope of the simultaneity slices of that frame is simply V (recall that light cone runs at 45° and bisects the angle between the space and time axes in every frame so the slope of the time axis is the reciprocal of the slope of the space axis, i.e. of the simultaneity slice.) The ratio of the mass of A to that of B in the rest frame of B is therefore $\dfrac{1+V^2}{1-V^2} : 1$. According to this frame, A, which has the same rest mass as B, has a relativistic mass of $\dfrac{1+V^2}{1-V^2} M_B$. Since B is at rest in this frame, this gives us the exact degree of mass increase. The only problem is that the ratio is not expressed in terms of the velocity of A in this frame, but rather in terms of the speeds of both A and B in the rest frame of the center of mass. To get the usual expression for the relativistic mass increase, we need to first find the speed of A in B's rest frame. Since the center of mass is moving at speed V relative to A, and B is moving at V relative to the center of mass, we can use the Relativistic velocity addition formula to get the speed of A relative to B. It is a consequence of the Lorentz transformations that if object 1 has velocity V_1 in frame of reference 1, and if frame of reference 1 is moving at velocity V_2 according to frame of reference 2, then the velocity of object 1 in frame of reference 2 is *not* $(V_1 + V_2)$, as the Galilean transformation would lead us to expect. If the velocities did add in this simple way, then a particle traveling at 99 percent the speed of light in Frame 1 would have to travel at more than the speed of light in a frame in which Frame 1 is itself moving at 99 percent of the speed of light. In Relativity, if object 1 is traveling at V_1 in Frame 1, and Frame 1 is traveling at V_2 in Frame 2, then object 1 travels at $\dfrac{V_1 + V_2}{1 + V_1 V_2}$ (all velocities are expressed in natural units, in which the speed of light is 1). Since A travels at velocity $-V$ in the rest frame of the center of mass, and the rest frame of the center of mass travels at $-V$ in the rest frame of particle B, particle A travels $W = -\dfrac{2V}{1+V^2}$ in the rest frame of B. The increase in A's mass expressed in terms of its velocity W is just a matter of some manipulation:

$$\frac{1+V^2}{1-V^2} = \left(\frac{1-V^2}{1+V^2}\right)^{-1} = \left(\frac{(1-V^2)^2}{(1+V^2)^2}\right)^{-\frac{1}{2}} =$$

$$\left(\frac{1-2V^2+V^4}{1+2V^2+V^4}\right)^{-\frac{1}{2}} = \left(1-\frac{4V^2}{1+2V^2+V^4}\right)^{-\frac{1}{2}} =$$

$$(1-W^2)^{-\frac{1}{2}}.$$

Since we have been using natural units, our final result is that if the rest mass of a particle is m_0, then the mass attributed to it in a frame of reference in which it is moving at velocity v is $m_0(1-v^2/c^2)^{-\frac{1}{2}}$: the mass scales by the same factor γ which appears in the Lorentz transformation.

In a given frame, then, as the velocity of a particle increases so does its mass. Furthermore, as the velocity approaches the speed of light, the mass approaches infinity, and hence it requires unbounded amounts of energy to get the particle to go faster. Given that only finite amounts of energy are available, it would follow that *no particle which travels below the speed of light can ever be accelerated to, or beyond, that speed.* So the speed of light does serve as a limit on the velocities of particles which start out traveling at sublight speeds.

As usual, it is therapeutic to look at this phenomenon from other frames of reference. When we try to accelerate particles in, say, a linear accelerator we find that the closer they get to the speed of light the heavier, and hence the harder to accelerate, they are. We can get electrons going 99.9 percent of c, but can't quite push them that extra little bit. But from the point of view of other frames we have been doing the wrong thing. There is a perfectly legitimate frame of reference in which the electrons come to be at rest after passing through the accelerator. From that point of view, we took electrons that were already going at nearly the speed of light and slowed them down. And from the point of view of the electron (or at least of the inertial frame in which it, at any given moment, is at rest) no progress at all is made toward "catching up" with light. Light, as always, speeds away at just the same rate. Looked at this way, the problem of surpassing light speed is not one of increasing mass, but more a sort of conceptual impossibility, like trying to reach your own horizon. No matter which way you go, it recedes apace and always keeps a constant distance away.

So all the usual matter we are familiar with, electrons and protons, quarks and pions, cannot travel faster than light. And all massless particles are constrained to travel at light speed. If the only means of communication or causal influence employ such particles, we would already have a fundamental conflict between the predictions of quantum mechanics and Relativity. It is true that the only means of causal influence we know of employ these particles or others which cannot break the light barrier. But all we can infer from this is that the photons, however they communicate, do not do so by sending electrons or other normal matter between them.

It is worthwhile noting how significant is the fact that exchange of normal matter and light cannot account for the correlations between particles. In "Quantum Mysteries for Anyone" (1981), David Mermin set out the correlated pair experiment emphasizing that the source and the detectors must all be unconnected to one another:

> The device has three unconnected parts. The unconnectedness lies near the heart of the conundrum, but I shall set it aside in favor of a few simple practical assertions. There are neither mechanical connections (pipes, rods, strings, wires) nor electromagnetic connections (radio, radar, telephone, or light signals) nor any other relevant connections. Irrelevant connections may be hard to avoid. All three parts might, for example, sit atop a single table. There is nothing in the design of the parts, however, that takes advantage of such connections to signal from one to another, for example, by inducing and detecting vibrations in the table top. (p. 398)

Mermin goes on to assure us that no matter how hard we look, we will not find any hidden gadgets inside the parts that allow them to communicate.

We can now see that the problem is both simpler and more perplexing than Mermin suggests. *It wouldn't matter* if the detectors were crammed full of radio sets or radar, or connected by pipes or strings, or attuned to vibrations on the table. In Aspect's set-up, none of that would help. None of those methods, indeed no known method, could get the needed information across in time to make a difference. Pipes and strings and vibrations in tables and radio waves are all constrained by relativistic considerations: none can go faster than light.

So we have derived a real constraint from relativistic considerations: particles which, at some time, travel below the speed of light cannot be accelerated above that speed. We have no conflict with the quantum non-locality, though, since there is no necessity, nor indeed any suggestion, that any such particles are used to secure the quantum connection. In fact, quantum theory requires that there be no matter or energy transfer between the correlated particles, at least in the following sense. After the paired photons or electrons have separated, the so-called interaction Hamiltonian between the particles decays to zero. The Hamiltonian governs the time evolution of the system,[2] and is intimately connected to its energy. If the interaction part of the Hamiltonian goes to zero, the evolution of the two wings decouples in certain ways. It follows that (1) no manipulation on one wing (where a manipulation is represented by a change in the Hamiltonian) can effect the expectation value of the energy on the other. As a consequence (2) if the particles have definite energies after they separate (if they are in eigenstates of the separate Hamiltonians) then nothing done on one side can change the energy on the other. In this sense, no energy or matter can be transmitted between the wings.[3] Quantum correlations do not require any superluminal matter or energy transfer. To the extent that Relativity forbids such transfer, all is still well.

Of course, having argued that the quantum correlations demand that some influence travel faster than light, this leaves us yet with a puzzle. If Relativity forbids energy transfer at superluminal speeds, then the influence, whatever it is, does not involve energy being sent from one wing to the other. Since all known means of communication utilize energy transfer, the quantum connection indicates some fundamentally new sort of phenomenon. We have purchased consistency with Relativity, at least in this respect, but at the price of enshrouding the nature of the quantum phenomena in yet more mystery.

Furthermore, the reader will doubtless have noticed that the supposed constraint on energy or matter transfer was not properly derived from relativistic considerations. All we have shown is that the sort of matter or energy which travels at or below the speed of light cannot be induced to transcend those bounds. But showing that things cannot be accelerated beyond the light barrier does not show that there can be no superluminal particles, only that they cannot be the same sorts of particles we are familiar with. Let us therefore consider what such a superluminal particle would be like.

Tachyons

The Lorentz transformations and the relativistic mass increase do not *per se* rule out superluminal particles (tachyons), but only prohibit the acceleration of particles through the light barrier. Tachyons must be born traveling faster than light and (as we will see) cannot be slowed to sublight speeds. But how are we to determine the properties of tachyons?

The easiest strategy is to simply insert superluminal velocities into the transformation equations we have already derived. This technique is of some use, but must be applied with care. We must, in particular, distinguish the project of describing particles in a given reference frame from that of generating new reference frames. The Lorentz transformations were concerned with the latter problem, and plugging superluminal velocities into them yields pointlessly confusing results.

The fundamental problem arises from the factor $\gamma = (1 - v^2/c^2)^{-\frac{1}{2}}$ Recalling the standard Lorentz transformation:

$$t' = \gamma \left(t - vx/c^2 \right)$$

$$x' = \gamma \left(x - vt \right)$$

$$y' = y$$

$$z' = z,$$

we find the disquieting result that in the tachyon's frame of reference both time and one of the spatial coordinates become imaginary (since $v > c$, γ is imaginary). What could this mean? Does the tachyon somehow travel in imaginary space and time?

This problem is, however, illusory. The "tachyon frame" that we generate by plugging in $v > c$ is not really a fundamentally new kind of reference frame at all. The coordinate hyperplane families of $t' = $ constant, $x' = $ constant, $y' = $ constant, and $z' = $ constant are identical to the families of hyperplanes in the reference frame associated with the sublight velocity c^2/v, as can be checked by calculation. The "tachyon frame" can be generated from the subluminal c^2/v frame by the simple substitution:

$$x'_{tachyon} = (-ic)t_{subluminal}$$

$$t'_{tachyon} = (-i/c)x_{subluminal}.$$

The scaling factors $-ic$ and $-i/c$ are put in to keep the interval $c^2 \, \Delta t^2 - \Delta x^2 - \Delta y^2 - \Delta z^2$ invariant under the transformation. Thus

$$c^2 \, \Delta t'^2 - \Delta x'^2 - \Delta y'^2 - \Delta z'^2$$

$$= c^2 \left(-1/c^2 \right) \Delta x^2 - \left(-c^2 \right) \Delta t^2 - \Delta y^2 - \Delta z^2$$

$$= c^2 \, \Delta t^2 - \Delta x^2 - \Delta y^2 - \Delta z^2.$$

So all the tachyon frame does is give the name "time" to a space coordinate in a subluminal frame and the name "space" to a time coordinate, with the compensating imaginary numbers added to keep the interval invariant.

There is no physical justification for regarding the tachyon frame's "time" as a time coordinate. We have no physical notion of a tachyonic clock since no normal clock mechanism could be accelerated to superluminal speed. Furthermore, if one adopts the tachyon frame's "time" and "space" labels, one gets the unsettling result that two events could be at both zero temporal and zero spatial separation from each other yet be distinct events. (Consider the locus of points where the $x = 0$ hyperplane of the associated subluminal frame intersects the light cone of the origin. In the tachyon frame,

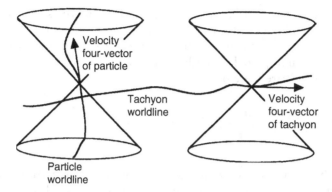

Figure 3.5 Velocity Four-vectors

these events all occur at $t' = 0$, and the invariant interval between them and the origin is zero since they lie on the light cone of the origin. Hence both $\Delta t'$ and $\sqrt{(\Delta x'^2 + \Delta y'^2 + \Delta z'^2)}$ are zero.) The most natural response to these worries is just to forgo the notion of a tachyonic frame at all. In the sequel we will consider only the problem of how tachyons would be described in the various familiar sublight frames of reference, leaving the question of "how things would look to the tachyon" to mystics and theologians.

The fundamental quantities that we use to describe relativistic particles are the four-velocity and four-momentum (or energy-momentum) vectors. It will be easiest to present these quantities graphically. If we trace out the trajectory of a particle in space-time, the four-velocity of the particle at any given time can be represented by the tangent to the trajectory (see figure 3.5). Tachyons differ from normal particles in that their four-velocity vectors are always space-like, pointing outside the light cone, while the four-velocities of normal particles are time-like, remaining inside the light cone.

There is, however, a slight embarrassment about the four-velocity of a tachyon. The formal definition of the velocity four-vector **U** in a given frame of reference (t, x, y, z) is:

$$U = (dt/d\tau, dx/d\tau, dy/d\tau, dz/d\tau), \text{ where}$$

$$d\tau = \sqrt{dt^2 - (dx^2 + dy^2 + dz^2)/c^2}$$

is the differential of the *proper time* of the particle. For a tachyon, this differential proper time is imaginary, not real. Fortunately, the important dynamic quantity is not the velocity four-vector but the momentum four-vector $P = m_0 U$. By assigning the tachyon an imaginary rest mass, the dynamical quantities all become real again. The tachyon's imaginary rest mass is acceptable since the tachyon can never be brought to rest.

The components of the momentum four-vector in a given frame of reference are easily interpreted using the fact that $dt/d\tau = (1 - v^2 / c^2)^{-\frac{1}{2}} = \gamma$. The components of **P** are therefore:

$$\left(\gamma m_0, \gamma m_0 \frac{dx}{dt}, \gamma m_0 \frac{dy}{dt}, \gamma m_0 \frac{dz}{dt}\right).$$

The last three components are just the momentum of the particle according to this frame: its velocity multiplied by its relativistic mass. The first component, the time-like component, is proportional to the relativistic mass itself. By the relativistic equation $E = mc^2$, this component is also proportional to the energy of the particle. The energy of a tachyon *decreases* as it goes faster (in a given frame) since γ approaches zero. The energy reaches zero when the velocity is infinite in that frame and the worldline of the tachyon lies within a hypersurface of simultaneity. Conversely, the energy increases as the particle is slowed closer to light speed. It would take infinite energy to slow a tachyon below the light barrier just as it takes infinite energy to boost a regular particle above it.

We now have enough information to explore the properties of tachyons. In order to do us any good in explaining the violation of Bell's inequality, the tachyons must serve as the medium of interaction between ordinary particles. Modeling such an interaction as a tachyon traveling from one particle to another we get the situation of figure 3.6 (using a two-dimensional spacetime). The simultaneity slices of one reference frame have been included. In this reference frame the tachyon is emitted by particle A just before t = 3 and is absorbed by particle B after t = 4. Energy and momentum are conserved at both the emission and absorption events. The particle A recoils after the emission and particle B is knocked off its old trajectory by the absorption. Superluminal energy transfer between the particles is achieved.

Interpretative problems arise, however, when we switch to new frames of reference. Since the emission and absorption events are space-like separated, there will be frames in which their time order is the opposite of that just given. The simultaneity slices of such a frame are given in figure 3.7. If we try to maintain our first interpretation of this interaction we are now faced with several difficulties. A emits the tachyon at about t' = 4 and it is absorbed by B at t' = 3! Our tachyon appears to be traveling not only faster than light but backward in time. A second difficulty is generated along with this first. The four-momentum vector of the tachyon is now pointed in the negative time direction, having dipped below the simultaneity slices. Since the energy of a particle is proportional to the time component of its momentum four-vector,

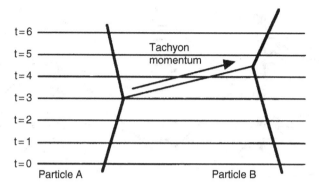

Figure 3.6   Tachyon Exchange in one Reference Frame

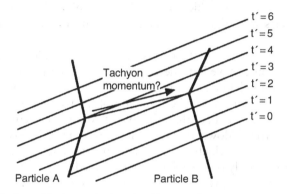

Figure 3.7   Tachyon Exchange in Another Frame

this means that the tachyon now has negative energy. Such negative energy particles are an embarrassment since they would introduce instabilities into any system: there is clearly no limit on the number of negative energy particles a system could emit.

The standard response to these worries is to introduce a *reinterpretation principle*.[4] When specifying the direction of the momentum four-vector of the tachyon we made an arbitrary choice: we could just as well have put the arrow of the vector on the other end, having it point in the opposite direction (this freedom derives from the fact that we could have defined $d\tau$ as either the positive or negative square root of $dt^2 - (dx^2 + dy^2 + dz^2)/c^2$). If we make this choice in the t' frame, then the interaction is redescribed as the emission of a positive-energy tachyon by B at $t' = 3$ and its absorption by A at $t' = 4$. The emission of negative-energy tachyons can always be redescribed as the absorption of positive-energy ones. Although which

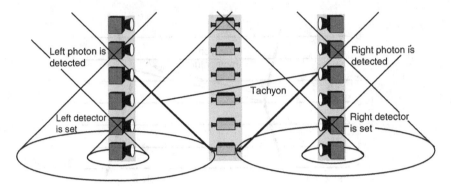

Figure 3.8   A Tachyonic Signal

particle is the source and which the receiver becomes relative to a reference frame, all frames agree that particles travel forward in time and always have positive energy.

The relativistic theory of tachyons can thus be made consistent. Furthermore, tachyons, as we have described them, are perfectly localized particles. They occupy definite continuous trajectories in space-time. Einstein's characterization of our physical conception of the world cited at the beginning of chapter 1 is satisfied: the principle of contiguous action holds. The world can be broken down into separately existing infinitesimal volumes of space-time, each of which has its own intrinsic state and each of which directly influences only others which are in immediate propinquity.

Unfortunately, tachyons cannot be pressed into service to explain the violation of Bell's inequality. For to do the job, the particles must carry the right sort of information and must have the right kind of effects. But these requirements run afoul of the reinterpretation principle.

Let's add a tachyon to our picture of the Aspect experiment. When the right-hand photon is observed it sends a tachyon to the left-hand one. The tachyon causes the left-hand photon to respond appropriately when observed (figure 3.8). In some frames of reference this story works. If in a given frame the detection event on the right occurs before the event where the tachyon meets the photon on the left, then the sequence is interpreted as a tachyon emitted by the photon when observed and absorbed by its partner in time to have an effect on its behavior. But in other frames the tachyon is perfectly useless. In the frame whose simultaneity slices are horizontal in this diagram, the tachyon is emitted by the left-hand photon before the latter is observed. It is therefore unable to carry any useful information to the partner, where it arrives at the observation event.

In fact, the problems run deeper than this. If the tachyon is to play its role even in the frames where the time order is correct, the nature (or existence) of the tachyon must somehow depend on the sort of observation carried out on the right. Let's suppose that there are two types of tachyons: positive and negative. Then we can imagine that only positive tachyons are sent if the polarizer on the right is in one position, only negative tachyons if it is in the other. Or we could imagine that a tachyon will only be sent if the setting is one way, none will be sent otherwise. In any case, there must be some strong correlation between the setting on the right and the type (or existence) of absorption event on the left. For if the same type of tachyon is sent no matter how the polarizer is set, the left-hand photon will gain no useful information by its arrival.

But looked at in other frames of reference the situation is now more puzzling than ever. In those frames, the left-hand photons manage to emit, say, positive tachyons exactly when the right-hand polarizer is set a certain way. These sorts of distant correlations among space-like separated events are exactly the problem we began with. We are now multiplying our difficulties rather than resolving them.

Advocates of the reinterpretation principle may try to maintain that the description offered in the last paragraph is incorrect. Rather than the photon being otiose, one should regard it as a *cause of the setting of the detector*. If this were so then the correlation between the nature of the particle and the setting would not be mysterious at all. The intuition behind this suggestion is that causal connections are a matter of correlations between events, and the further judgment about which is the cause and which the effect depends merely on which is precedent in time. A Lorentz transformation which inverts the time order of two connected events therefore would also invert the assignment of the labels "cause" and "effect."

But such a response cannot be maintained in this case. First of all, the setting of the detector can easily be arranged to be completely determined independently of any "incoming" tachyons. We might, for example, preset a computer to change the setting in a patterned way. The tachyon could not, in such a case, be interpreted as the determining cause of the setting. Skeptics could be convinced of this by running the experiment in two modes: one with only the computer on and no photon pairs being produced, the other with only the photon pairs and no computer. The polarizer setting would change only in the first mode, demonstrating that the computer alone caused the setting.

It is easy to be misled here by a bad analogy. Newtonian mechanics is time-reversal invariant: if a movie of a collision between Newtonian billiard balls is shown backwards it will depict a possible Newtonian sequence of events. In the backwards showing, events which were originally causes come to be effects. One may therefore conclude that any transformation which changes the time order of events must thereby demand a revision of their causal connections. But unlike the Newtonian case, the Lorentz transformation does not result in *global* time inversion. Only certain pairs of space-like separated events have their order of precedence in time changed. There are many things about which both reference frames still agree. In the example sketched above, they would both agree in their analysis of the operation of the computer and of the connection of the computer with the polarizer, and would therefore agree that the computer is the complete cause of the polarizer setting. Any correlation of that setting with the production of distant tachyons would be truly mysterious.

Tachyon interactions appear to be possible, but they can be reconciled with the reinterpretation principle only if the interaction is appropriately symmetric. In our first example, this condition is met: the deflection of particle A can equally well be understood as the recoil due to emission of the tachyon or as the effect of its absorption, and the same is true of B. Nothing in the diagram of the interaction singles out one or the other particle as source or receiver. These sorts of interaction, involving the exchange of so-called virtual particles, are common coin in quantum field theory, being the elements from which Feynman's path-integral formalism is constructed. For this reason it is sometimes tempting to see the relativity of cause and effect as already well established in quantum theory.

But as soon as the tachyons are fitted out to do the work we require of them the essential symmetry disappears. The nature of the tachyons must be correlated to one or the other (or both) of the polarizer settings. And in some frames of reference this correlation will be inexplicable since the tachyon will be created far from the polarizer and before it is set.

The lesson of the tachyons will recur throughout our examination. In the case of symmetric interactions, one is able to preserve the notion that *the future is determined only by the past*. Different frames of reference may disagree about time orders of events, and hence about what constitutes the past of any given event, but still each frame is able to construct a story consistent with its own temporal structure. But explaining the quantum correlation seems to require breaking of that symmetry. Some events are identified as causes and others as effects in a way that cannot be reversed. This raises two specters. The first is that backward causation, like the motion of particles backward in time, may introduce insurmountable problems into our physical theory. The other is that the physical structure may, at base, no longer be Lorentz invariant. If there is some unique frame of reference in which causes always precede effects then that frame has a claim to preferred status. If Nature transacts her business in one coordinate system rather than any other, the Principle of Relativity is more illusion than insight.

Notes

1 The notion of a *causal influence* is notoriously difficult to explicate. We will take up this issue with more precision in chapter 5.

2 Except (an important exception!) during wave collapse, if any.

3 It is a more difficult matter to say what happens if the respective particles are not in eigenstates of their Hamiltonians and the system undergoes a collapse. In some sense, each particle has an indefinite energy before the collapse and a definite energy after. This is not obviously a case of energy *transfer* since the energy was not previously "located" on one wing or the other. But neither is it obviously not a case of transfer.

4 See, for example, Feinberg (1967), where all of these problem are discussed in detail.

4

Controlling the Connection: Signals

Transmission of energy faster than light presents a subtle threat to our physical theories, as we saw in the last chapter. If we are unable always to invoke the reinterpretation principle, some frames of reference will have to tolerate negative energy particles which move backwards in time. Such negative energy states could be a source of radical instability, since a system might be able to boost its energy indefinitely by emission of such particles.[1] But the most common characterization of the relativistic constraint does not advert to energy transmission or dynamical instability. Rather, relativity is often portrayed as implying, or even being derived from, a constraint on the speed with which *signals* can be sent.

The association of Relativity with claims about signals is so common, and is so deeply enmeshed in the history of the theory, that it is easy to overlook how startling it is. Energy, momentum, mass, charge, and so on, are the quantities which appear in the fundamental equations of physics. But by what principle are we to recognize a particular physical transaction as a case of signaling? Without clear criteria for identifying signals we face the possibility of having formulated a basic physical principle using terms which are vague or obscure.[2] So our first project is to explicate the very notion of a signal in some exact terms.

As usually conceived, signaling is a process by means of which humans (or other creatures or things appropriately like humans) can communicate with each other. When we characterize one system as a transmitter and another as a receiver of a signal, we require that a *controllable* aspect of the transmitter be correlated with some *observable* aspect of the receiver. The notion of a signal is thereby doubly anthropocentric: it depends on a

Quantum Non-Locality and Relativity: Metaphysical Intimations of Modern Physics,
Third Edition. Tim Maudlin.
© 2011 Tim Maudlin. Published 2011 by Blackwell Publishing Ltd.

prior specification of what the sender can freely manipulate and what the observer can see. As Bell notes, "to answer this we need at least a schematic theory of what *we* can do, a fragment of a theory of human beings" (1987, p. 60). If the observable aspect of the receiver is so correlated, then from what she sees the recipient can more or less reliably infer what the sender is doing. The reliability of the inference depends on the strength of the correlation. In the ideal case the correlation is perfect and the message can be decoded with certainty. An imperfect but non-zero correlation can also be used ("signals with noise") although the inference to the state of the transmitter will sometimes fail. If there is no nomic correlation at all between transmitter and receiver then no message can be sent.

In our schematic version of the correlated-pair experiment, the specification of what is assumed to be controllable and observable by us is easily made: we can freely decide which question (if any) to ask and we can tell what answer is given. We cannot force the particle to give a particular answer, nor can we observe anything about its state apart from the results of our experiment. In concrete terms, we can decide how the polarizers are to be set and can observe only whether or not a photon passes a given polarizer. So the question of superluminal signals comes to this: can manipulations of the polarizer on one wing of the experiment produce a noticeable effect on the other, given that the Bell inequalities are violated?

The short answer is that the violation of the inequalities does not *per se* imply any ability to send signals. It is easiest to see this by construction. Here is a mechanism by which the predictions of quantum mechanics may be recovered and Bell's inequality violated. When one of the particles gets to its polarizer, it flips a perfectly random fair coin to decide whether to pass or not. It then communicates both the polarizer setting and the result of the coin flip to its partner. The partner, when measured, adjusts its responses to accord with the quantum correlations. If its polarizer is set in the same direction as its partner's it always agrees with the partner, if the polarizer is off by 30° it agrees 75 percent of the time (whenever, say, either of two coins comes up heads), if the polarizer is off by 60° it agrees 25 percent of the time (whenever both coins come up heads). By construction the quantum correlations are recovered. But it is also obvious that no manipulation of polarizers on one side will make a noticeable difference on the other. For no matter how the polarizers are set, the probability at the beginning of the experiment that a photon will pass a given polarizer is always 50 percent. For the "first" photon, this is due to the decision being based on the coin flip. For the second, it is because what counts as agreeing or disagreeing with its partner similarly depends on the result of that flip. If that first decision is truly random, then all one will see on either wing of the experiment, no matter how the distant polarizer is set, is a random sequence of passage and absorption of photons.

It should be noted that this model of how the correlations are generated is not merely fanciful. It is, in fact, the standard account of quantum mechanics with wave collapse. One obvious problem is how the determination is made that one of the photons is the "first" and the other the "second." In non-relativistic quantum mechanics time order of the observations is used: the one which gets to its polarizer first is "first." Our problem, of course, is that in a relativistic context this expedient is not available.

Both the restrictions on what we can control and those on what we can observe are crucial in eliminating the possibility of signaling. If we could force a photon to either pass or be absorbed then we could signal by setting our polarizers in the same direction and modulating the behavior of the distant photon by controlling the local one. Or again, if we could directly inspect the state of a photon, to see whether it is disposed to certainly pass (or certainly be absorbed) if measured in some direction then we could know whether a measurement in the corresponding direction had been carried out on the distant wing. In this case our colleague could send a message by manipulating his polarizer.

There is yet another possibility for using the experimental set-up to send signals, a possibility which can be illustrated by imagining a very different underlying mechanism. Our first picture employed irreducibly stochastic processes at the measurement events, but violations of the inequality can also be generated in purely deterministic settings. Again suppose that when both photons are observed some principle determines one of the measurements as "first." Then the decision about how each photon will react in every circumstance can be already made at the source. The state of each photon could contain instructions such as: "If measured first and the polarizer is set at 0° then pass and send a message to the other photon" or "If a message is received saying the other photon was measured at 60° then pass if measured at 30° and do not pass if measured at 0°." This scenario corresponds to settling all the results in all possible experimental configurations at the source, thus leaving to be sent only the information about how the distant polarizer is set. A well-chosen mixture of such sets of instructions could reliably produce violations of Bell's inequality. Of course, some of the instructions must be conditional, dependent on the nature of the distant measurement: the impossibility of using categorical instructions that refer only to the local detector was the conclusion of Bell's proof. (Again, this is not a fanciful model: it is, in essence, David Bohm's picture of the reality underlying quantum phenomena. We will discuss Bohm in detail below (appendix B).)

In this scenario messages could be sent if the state of the photons at the source could be either observed or controlled. Since most of the instruction sets will contain conditional instructions, the results of some observations on one side must depend on the sort of measurement carried out on the other. If the instructions can be known at the source (by either producing or observing them) then they can be known to the experimenters, who could agree on a scheme to use the conditional outcomes to send signals. If, for example, it is known that whether or not the right photon will pass a polarizer set at 30° depends on the sort of observation carried out on the left, then the experimenter on the left is free to control the outcome on the right, and can use that freedom to transmit information.

In the first scenario the state of the photons at the source contains no hidden information: all of the pairs are created in precisely the same state with the differences in outcomes for different pairs deriving from the stochastic element. Observation of the particles at the source could not provide useful information since only one state is ever created, and for the same reason there is nothing at the source to control. The second scheme is a deterministic "hidden-variables" model in which all outcomes are decided by antecedent states. In order to avoid possibilities for signaling the hidden variables must to some extent remain hidden and out of reach of experimental control. So long as the observer's powers of observation and manipulation are circumscribed in these ways the correlated pairs cannot be used for sending signals. Many variations on these schemes can be imagined, but the general method of avoiding signaling for each case should be clear.

If all that Relativity demands is the impossibility of superluminal signaling then the violation of Bell's inequalities does not *per se* pose any problem. The inequalities could be violated in a way that permits signaling, but they need not be. And so far as we know, so far as our theories inform us about what can be controlled and observed, faster-than-light telegraphs cannot be built. Or, as it is often said, there is no Bell telephone.

This leaves us with the problem of connecting Relativity to the constraint on signaling. As with superluminal energy transport, it is not immediately obvious why the relativistic space-time structure should forbid such possibilities, or why, if it does, that prohibition should be regarded as the very essence of Relativity. We shall turn to this question presently, pausing first for a more detailed look at how the constraint on signaling is woven into the structure of quantum theory.

A Slightly Technical Interlude[3]

We mentioned above that what implications a physical theory has about signaling may not be immediately obvious from its formalism: in addition to the equations which govern the physical state we must obtain a specification of what can be controlled and what can be observed by us. This left us with a position both vague and seemingly anthropocentric. We can't do any better than this until we have a theory in hand. We can do better with quantum theory as it is usually formulated.

The orthodox quantum description of a system involves three elements. First, the system is assigned a state, represented by a vector in a Hilbert space. Second, the Hamiltonian of the system determines how that vector will evolve in time so long as no measurements are made on the system.[4] Finally, when a measurement or observation is made, the quantum state is changed discontinuously ("wave collapse") in a way that reflects the outcome of the measurement. In particular, a quantum state which does not represent the system as having a definite value of the observable magnitude collapses to one which definitely has the observed value.

In order to apply our analysis of signaling to the quantum formalism we must identify the observable and controllable aspects of a physical system. There is a precise mathematical characterization of the former: if the states of the system are represented by vectors (or rays) in a Hilbert space, the observables correspond to Hermetian operators on that space. The question of what is controllable is more problematic. With regard to influencing the particle on one wing by changing the Hamiltonian of the system, and thereby changing its time evolution, Bell supposes that any change in the system due to a manipulation performed on some quantity b will give rise to the following change in the Hamiltonian H:

$$\delta \int dt H = B\, \delta b,$$

where B is an operator localized in the region where the manipulation is performed (see Bell 1987, p. 61). With regard to non-Hamiltonian time evolution, we suppose that an observation on one wing of the experiment can result in a collapse of the quantum state to a state with a definite value of the observable. *Which* definite state results is *not* under our control: it is here that the irreducibly stochastic element in the orthodox interpretation of quantum mechanics appears.

We are now faced with two questions. First, can a change in the Hamiltonian such as Bell describes produce a noticeable change on the distant wing? Second, can wave collapse induced by an observation on one wing produce such a change on the other? The answers to both of these questions is "No," but the analysis in each case is rather different.

To send a signal we must produce an observable change on the distant wing. Suppose the observable feature of the distant wing we intend to influence is represented by the Hermetian operator A. Will the change in the Hamiltonian due to the manipulation of b produce any change in the expectation values, i.e. the statistical predictions, for observed values of A? If the change in the Hamiltonian is as above, then in general the change in A as a function of the change in b is

$$\delta A/\delta b = [A, -(1/h)B],$$

where the square brackets represent the commutator of the operators. We have assumed that A and B represent operators localized to the respective sides of the experiment, i.e., operators which are functions only of variables which pertain to those regions. Now it is a *postulate of relativistic quantum field theory* that the commutator between any observables which are space-like separated is zero.[5] These so-called *Equal Time Commutation Relations* guarantee that the expectation value of A will be unchanged by manipulations of b, and so no signal can be sent in this way (ibid., p. 61).

The case of wave collapse is more subtle. It is most easily illustrated by considering a special case, such as our two correlated photons. When the pair is created, quantum theory describes it using a so-called *entangled* state, which may be written as:

$$1 / \sqrt{2}(| \text{ pass } >_{R0°} | \text{ pass } >_{L0°} + | \text{absorb} >_{R0°} | \text{absorb} >_{L0°}),$$

where |pass $>_{R0°}$ represents the state in which the photon will definitely pass a polarizer oriented at 0° and so on. The choice of 0° here is arbitrary: the same state could be described in the same way using any other polarizer angle. When the pair is in the entangled state the probability of either photon passing a detector is 50 percent. The state is a superposition of a state in which both will pass and one in which both will be absorbed if measured in the same direction; it does not determine which result will occur. But since in each of the two superposed states the photons *agree* with one another it is determined that they will be perfectly correlated if identical measurements are made on both sides.

Suppose that the photon is measured at 0° and passes the polarizer. The pair will now be assigned the state |pass $>_{R0°}$ |pass $>_{L0°}$. The statistics of this state with regard to observations on the left are quite different from those of the entangled state. There is now a 100 percent probability that the left-hand photon would pass at 0° or, more generally, a probability of $\cos^2 \theta$ that it will pass at angle θ. If we regard wave collapse as a change in the physical state of the particles, the measurement on the right has created a quite distinct state of affairs on the left with respect to the probability of various observable events happening. Why couldn't this change be used to transmit information?

The solution relies on the fact that although we can decide to perform an observation on the right, and so cause a collapse, we cannot control whether the state collapses to |pass $>_{R0°}$ |pass $>_{L0°}$ or to |absorb $>_{R0°}$ |absorb $>_{L0°}$: each will occur 50 percent of the time. So if we start with a large ensemble of correlated pairs in the entangled state and measure all of the photons at 0°, the ensemble will be changed from having all of the pairs in the entangled state to having (roughly) half in |pass $>_{R0°}$ |pass $>_{L0°}$ and half in |absorb $>_{R0°}$ |absorb $>_{L0°}$. That is, from being in a pure state the ensemble goes into a mixture of $\frac{1}{2}$ | pass $>_{R0^\circ}$ | pass $>_{L0^\circ} + \frac{1}{2}$ | absorb $>_{R0^\circ}$ | absorb $>_{L0^\circ}$. This mixture is not statistically identical to the pure state in all respects, but *it is statistically identical with respect to observables which pertain solely to one side or the other*.

More precisely, if we begin with a description of a system as a vector in a Hilbert space which is the tensor product of several different Hilbert spaces which are associated with different parts of the system, then the measurement of an observable associated purely with one of the subspaces will create a mixture which has unchanged statistical properties with respect to observables associated purely with other subspaces (see Redhead 1987, pp. 51–9).

At the beginning of the experiment we would assign particular probabilities to the results of different experiments which could be carried out on the right. If we now calculate the probabilities for those results conditional on the fact that a particular observation has been carried out on the left (with unknown result), they will be unchanged. So no observations carried out on the right can provide reason to believe a measurement has occurred on the left. Wave collapse, though it seems to alter the propensities for events to occur in distant regions, cannot be used to send signals since the way those propensities are changed cannot be controlled.

Bell's Theorem Again

In chapter 1 we proved a special case of Bell's theorem for an experimental arrangement which displays perfect correlation of results on the two wings when the polarization is measured in the same direction. We noted that similar inequalities could be proved for more general cases, but rested content with the one striking instance. We have now, however, reached a point of contact with philosophical analyses that embark from the more general case, and so pause to provide arguments similar to Bell's original proof. The particular proof offered below follows Abner Shimony's "An Exposition of Bell's Theorem" (Shimony 1990) and draws elements from Bell's proof in "The Theory of Local Beables" (reprinted in Bell 1987, pp. 52–62).

Consider any physical theory which explains the results of two distant measurements performed on a pair of particles by reference to the antecedent state of the particles. In particular, consider a theory which must explain the lawful, predictable correlation between the results of such measurements. Such a theory will ascribe to the pair some physical state, $k \in K$, which describes the condition of the particles as they leave the source. Assume that on the right-hand side we are free to choose to set our measuring device in any state $a \in A$ (A might include all orientations of the right polarizer) and

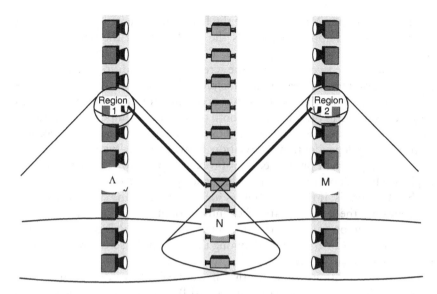

Figure 4.1   Space-time Diagram of Experiment on Pair

that the result of a measurement will be accorded a numerical value $s \in S$ (s might equal 1 if the photon passes and –1 if it is absorbed). Similarly, the left-hand detector can be set in a state $b \in B$ and will yield a result $t \in T$. Without loss of generality, we suppose that both s and t lie in the interval between –1 and 1 inclusive. k is a complete specification of the state of the pair, including whatever hidden or unknown factors one wishes to suppose. The theory might be either deterministic or stochastic; in the latter case, k would only fix the probability of getting particular outcomes for various measurements. But Bell supposed in his original paper that given a particular k, the probability of getting the joint result (s, t) if one measures a and b is prob(s/a, k) × prob(t/b, k), that is, he assumed that the joint probability for two results factors into the product of the probabilities for the results individually, once k is given.

The implications of and justification for this assumption have come under very close scrutiny, so it will be worthwhile to examine Bell's later formulation of the problem. Suppose that measurement events take place in space-like separated regions 1 and 2 as shown in figure 4.1. Let N represent the region of space-time which is the overlap of the backwards light cones of regions 1 and 2. Λ is the remainder of the past light cone of region 1 and M is the remainder of the past light cone of region 2. If we imagine our specific photon polarization experiment sketched into this general picture, region 1 contains the detection of the left-hand photon, region 2 the detection of the right-hand photon, Λ the setting of the left-hand apparatus, M the setting of the right-hand apparatus, and N the production of the photons at the source.

We now invoke two principles: the screening-off condition for conditionalizing on common causes and a principle of local causality. The screening-off condition asserts that once one has accounted for all events which could play a causal role in bringing about both of two spatiotemporally separated[6] events, the probabilities assigned to those events by a theory should become statistically independent. A correlation between event-type A and event-type B obtains when $prob(A \& B) \neq prob(A) \times prob(B)$. The correlation is explained by a set of common causes $\{C_i\}$ when $prob(A \& B/C_i) = prob(A/C_i) \times prob(B/C_i)$, that is, if the probabilities for A and B become statistically independent once the factors C_i are taken into account. (We take as a degenerate case of a common cause the case in which either of A or B is a cause of the other. If either A or B is counted among the C_i then the screening-off condition is trivially satisfied.)

Our first condition just states that if we have taken into account every event which could serve to correlate A and B by causally influencing them both, then there should be no further predictable correlation between them. This condition is sometimes characterized as a demand for causal explanation for correlations, although that characterization is not quite apt. The screening-off condition allows that there be accidental correlations, due to pure chance, but requires that no such correlations be predicted by a theory. This is nearly tautologous – if a theory predicts a correlation, then that correlation cannot, according to the theory, be accidental. A nomic correlation is indicative of a causal connection – immediate or mediate – between the events, and is accounted for either by a direct causal link between them, or by a common cause of both.[7]

The principle of local causality states that all of the causes of an event must lie in its past light cone. In a locally causal world, the events in region 1 cannot directly influence the events in region 2 or vice versa. Therefore any correlations between such events must be explained by a common cause. And any common cause of both the events in region 1 and the events in region 2 must lie in N, in the intersection of their back light cones. Suppose that v is a specification of the entire physical state of the region N. In a locally causal world, v would specify everything which might give rise to correlations between the events in region 1 and 2. Therefore, the probabilities for events occurring in region 1 and region 2 would be statistically independent once one has conditionalized on v.

To fill out our case, suppose we are interested in calculating the joint probability for getting particular outcomes for the measurement events in regions 1 and 2. In a locally causal world, all the causally relevant factors must lie in the past light-cones of the regions. If λ is a complete specification of the physical state of the region Λ then the probability for getting a particular result in region 1 is completely determined by $\lambda + \nu$, and if μ is the physical state of M then the probability of a particular result in region 2 is determined by $\mu + \nu$. In particular, the prediction for region 1 would not be changed by adding information about what happens in M, nor the prediction for region 2 changed by information about the state of Λ. For the events in those regions, being space-like separated, could neither be causes nor effects of each other, and so could neither alter the probabilities for nor provide information about the outcome of the measurements in each other.

By the same reasoning, information about the events in region 2 could not effect the probabilities assigned in region 1 once the entire back light cone of region 1 has been accounted for. In a locally causal world, events in region 2 cannot be either causes or effects of events in region 1. Since the outcome t is a report about events in region 2, the prediction for region 1 would not be changed by knowing t once $\lambda + \nu$ has been taken into account.

So if we want to calculate the joint probability $\text{prob}(s,t/\lambda,\mu,\nu)$ we first note that $\text{prob}(s/\lambda,\mu,\nu) = \text{prob}(s/\lambda,\nu)$ and similarly $\text{prob}(t/\lambda,\mu,\nu) = \text{prob}(t/\mu,\nu)$. Finally, $\text{prob}(s,t/\lambda,\mu,\nu) = \text{prob}(s/\lambda,\mu,\nu,t) \times \text{prob}(t/\lambda,\mu,\nu) = \text{prob}(s/\lambda,\nu) \times \text{prob}(t/\mu,\nu)$ in a locally causal world since the conditionalization on ν would account for all possible common causes. Of course, in a locally causal *deterministic* world, each of the probabilities involved would be either 1 or 0, but no assumption of determinism is being made.[8] A locally causal world can also be irreducibly stochastic, but the causal structure would still render the probabilities for space-like separated events statistically independent once one has conditionalized each on the entire state of its past light cone.[9]

When analyzing the Bell correlations one usually stipulates that all of the causally relevant facts which are determined in the region N are to be represented in the state k of the particle pair. k includes the quantum state of the pair, but may also contain whatever "hidden" extra variable one wishes to postulate. One also generally assumes that the only physical event in Λ which is relevant to the outcome is the setting of the left-hand detector, a, and the only relevant event in M is the setting of the detector, b. Under these assumptions, $\text{prob}(s/\lambda,\nu)$ becomes $\text{prob}(s/a,k)$ and $\text{prob}(t/\mu,\nu)$ becomes $\text{prob}(t/b,k)$, so we finally recover the condition to be used in the proof.[10]

The *expectation value* for an observable relative to a given state is the long-term average one ought to get upon repeated measurements of that observable on that state. One calculates the expectation value by multiplying the various possible outcomes by their probabilities and summing, so that:

$$E(a, k) = \sum_{s \in S} \text{prob}(s/a,k)s$$

gives the long-term average for repeated measurements of observable a on state k. Similarly:

$$E(b, k) = \sum_{t \in T} \text{prob}(t/b, k)t.$$

Finally:

$$E(a, b, k) = \sum_{st} \text{prob}(s, t/a, b, k)s\ t$$

that is, the joint expectation value is the expected value of the product of the two observables, calculated by weighting each possible product by the likelihood of its occurrence. Since in a causally local world $\text{prob}(s,t/a,b,k) = \text{prob}(s/a,k) \times \text{prob}(t/b,k)$, in such a world $E(a, b, k) = E(a, k) \times E(b, k)$.

We now need to prove a general mathematical fact, namely, that if x', x'', y' and y'' are all numbers that lie between –1 and 1 (inclusive), then the quantity $x'y' + x'y'' + x''y' - x''y''$ lies between –2 and 2 (inclusive). The simplest proof of this fact begins with the observation that the quantity in question is a linear function of its four variables, and hence must take its extreme values at extreme values of its domain, i.e. at some of the corners of the domain where each of x', y', x'', and y'' are +1 or –1. Now each of the products in the sum must be either +1 or –1, so the corner value is an integer between +4 and –4. But the quantity can also be written $(x' + x'')(y' + y'') - 2x''y''$. Since $(x' + x'')$ and $(y' + y'')$ each is either +2, 0 or –2, the first term is either +4, 0, or –4. But the second term is +2 or –2, so the total cannot sum to ±4 or ±3.[11]

We assumed at the beginning that all the possible results of measurements would fall in the interval between +1 and –1, so the expectation values for these measurements must also fall into this interval. Consider the case where there are two possible measurements which can be performed on each side, a', b', a'', and b''. Letting x', y', x'', and y'' stand for the expectation values for these measurements, the fact proved above becomes

$$-2 \le E(a', b', k) + E(a', b'', k) + E(a'', b', k) - E(a'', b'', k) \le 2.$$

This result is for the particular state k of the particles, but over a long sequence of runs pairs in different states may be produced. Quantum mechanics assigns all of the pairs the same physical state, but we wish our proof to be perfectly general, so we allow K to contain any number of states. Still, there must be some distribution $\rho(k)$ which represents the frequency with which the various states are produced. We assume that the distribution $\rho(k)$ is statistically independent of the settings of the detectors since (1) the setting cannot affect k because none of the region N lies in the future light cone of the setting; and (2) k cannot affect the setting since we can have some deterministic system independent of k (and of what happens in N) do the setting.

The observed correlation over a long run of experiments where the states of the particle pairs are distributed according to $\rho(k)$ is obtained by integrating over the space K of states, weighting the integration by $\rho(k)$. For example, the expected observed value of the product of results when a′ and b′ are chosen is:

$$\int_K \delta k \; \rho(k) E(a', b', k).$$

If we call this weighted value $E_\rho(a', b')$ then the integration yields

$$-2 \leq E_\rho(a', b') + E_\rho(a', b'') + E_\rho(a'', b') - E_\rho(a'', b'') \leq 2$$

(using the normalization condition $\int_k \delta k \; \rho(k) = 1$).

Since the E_ρ are things which can actually be observed, this inequality can be directly tested. In our polarization case we need only assign the value +1 to a photon which passes the polarizer and −1 to one which is absorbed. Let a′ represent having the polarizer on the right be set at 30°, a″ represent that polarizer set at 0°, b′ represent the left-hand polarizer set at 30°, and b″ that polarizer set at 60°. Then the observed long-term averages are:

For a′, b′, the average is 1 since the photons always agree.

For a″, b′ and a′, b″, the average is 0.5 since the photons agree 75 percent of the time (+0.75) and disagree 25 percent of the time (−0.25) when the polarizers are misaligned by 30°.

For a″, b″, the average is −0.5 since the photons only agree 25 percent of the time when the polarizers are misaligned by 60°.

But $1 + 0.5 + 0.5 - (-0.5) = 2.5$, violating the inequality.

Our original proof was simpler since we used the perfect correlation of the photons to infer that any local laws governing their behavior must be deterministic. The proof just given makes no assumptions about perfect correlations at all, and explicitly addresses stochastic theories. The fundamental assumption was that the results on the two wings be statistically independent once one conditionalizes on the physical state of the intersection of the past light-cones of the measurements, since any local common causes must lie in that region. This condition of factorization has been subject to detailed analysis, consideration of which will bring us back to the question of signaling and Special Relativity.

What Does Factorization Signify?

Bell derived the factorization condition

$$\text{prob}(s,t/a,b,k) = \text{prob}(s/a,k) \times \text{prob}(t/b,k)$$

from the condition of local causality along with a principle of explaining correlations by common causes. Since factorization is the key to deriving the inequality it has been subject to careful analysis. The most influential account of its significance is due to Jon Jarrett (1984). Jarrett argues that the condition used is best understood as the conjunction of two different principles, each of which has its own physical significance. The first, which Jarrett calls "locality" and Shimony denominates "parameter independence," states that the result of a given measurement is statistically independent of the *setting* of the distant detector:

prob(s/a,b,k) = prob(s/a,k) and
prob(t/a,b,k) = prob(t/b,k).

The second, which Jarrett calls "completeness" and Shimony "outcome independence," holds that the probability of getting a result on one side is independent of the *result* on the distant wing if both the result *and* the distant setting are given, i.e.,

prob(s/a,b,k,t) = prob(s/a,b,k) and
prob(t/a,b,k,s) = prob(t/a,b,k).

If both of these conditions hold then:

prob(s,t/a,b,k) = prob(s/a,b,k,t) × prob(t/a,b,k)
= prob(s/a,b,k) × prob(t/a,b,k) (by outcome independence)
= prob(s/a,k) × prob(t/b,k) (by parameter independence),

and we again recover the factorization condition. Hence the experimental observation of violations of Bell's inequality shows that either parameter or outcome independence must fail.

What is one to make of this analysis? The first question which presents itself is whether the experimental results can tell us which of the two conditions is violated. Unfortunately, they do not. In some theories which recover the quantum predictions, parameter independence is violated, in others, outcome independence is. In the second mechanism presented at the beginning of this chapter, the deterministic model which corresponds to Bohm's theory, parameter independence fails. Indeed, in that theory prob(s/a,b,k) and prob(t/a,b,k) are both either 1 or 0: given the initial state and the settings of the detectors the results are fixed, so information about the distant results is redundant.[12] In the standard quantum mechanical theory with wave collapse parameter independence holds but outcome independence fails. In this case k is the quantum wave-function. If I know the wave-function, additional information about what is being measured in the distant apparatus will not affect my statistical predictions about what I will see here.

What significance do parameter and outcome independence have? It is tempting to suppose that since the relevant "parameter" is the distant detector setting and the "outcome" is the result (e.g. passage or absorption), each condition reflects a claim concerning the causal role of a particular event: parameter independence holding if the act of setting the distant device has no distant causal role; outcome independence holding if the measurement event itself has no distant effects. But this association of each condition with a different event is spurious. Indeed, the entire analysis is somewhat arbitrary, at least if understood as analyzing Bell's condition into a constraint on the causal role of the settings and a constraint on the causal role of the result.

It is true that Bell's factorization condition is logically equivalent to the conjunction of outcome and parameter independence. But it is equally logically equivalent to the conjunction of the following two principles:

P1 prob(s/a,k,t) = prob(s/a,k) and prob(t/b,k,s) = prob(t/b,k).
P2 prob(s/a,b,k,t) = prob(s/a,k,t) and prob(t/a,b,k,s) = prob(t/b,k,s).

(P1) says that the probability assigned to a given outcome on one side is not changed if one knows the result (for example, passed or absorbed) but not the setting on the other side. (P2) says that the probability of a given result on one side, assuming one already knows the result on the other, is not changed if one knows also the distant setting. One might very well call (P1) "outcome independence" and (P2) "parameter independence," since (P1) concerns conditionalizing on the distant outcome and (P2) on the distant setting. But now the verdict for a theory can be changed: orthodox quantum mechanics violates (P2) but not (P1). Factorization is no more correctly seen as the sum of outcome independence and parameter independence than as the sum of (P1) and (P2). And, as Bell saw, it is the failure of factorization alone which gives us the most startling result: no account can be given according to which all of the causes of every event lie in its past light cone.

If Shimony's terminology is slightly problematic, Jarrett's is positively misleading. Jarrett calls parameter independence "locality" and outcome independence "completeness."[13] He argues that a violation of locality would allow superluminal signals to be sent and further asserts without argument that the absence of superluminal signals is all that is required by Relativity. This yields the conclusion that

> since locality is contravened only on pain of a serious conflict with relativity theory (which is extraordinarily well-confirmed independently), it is appropriate to assign the blame to the completeness condition. (Jarrett 1984, p. 585)

Let us take these assertions in turn.

As we have seen, signaling requires particular abilities to control and to observe physical states. Since the "locality" condition is purely statistical, implying nothing about what can be controlled or observed, how does Jarrett intend to demonstrate that violation of it would allow signaling? He does so by simply presupposing what can be controlled or observed.

The idea is quite straightforward. It is fair to assume that we can control the apparatus settings and that we can observe the outcome of each measurement. Now suppose that "locality" is violated. Then for at least one state k either prob(s/a,b,k) ≠ prob(s/a,k) or prob(t/a,b,k) ≠ prob(t/b,k). Suppose the first inequality holds. Then there must be some b′ so that prob(s/a,b,k) ≠ prob(s/a,b′,k). If we can prepare a large ensemble of states known to be in state k then by setting the detector on the left to either b or b′ one could change the probability for observing the result s on the right. As the ensemble gets large, the probability approaches 1 that the proportion of s results seen on the right correlates with the setting on the left. Hence by manipulating the setting on the left, information can be sent to the right.

As is immediately obvious, Jarrett not only assumes that detector settings can be controlled and results observed, which is reasonable, but he also assumes that a large ensemble of particle pairs can be prepared which are known to be in state k.[14] Without this assumption none of his schemes for sending signals are practicable. Not surprisingly, Bohm's theory, which violates "locality," also yields principled restrictions on state preparations, restrictions which disallow the possibility of sending superluminal signals.

Jarrett's discussion of "locality" actually suggests that it is not superluminal *signals* but superluminal *causation* which he takes to be forbidden by Relativity:

> What locality *does* exclude, however, is the possibility that the preparation of either measuring device can exert a causal influence on the other subsystem so as to affect the probabilities for the possible outcomes of measurements performed on that other subsystem. Since these two events, the preparation of one measuring device in a given state and the measurement executed by the other measuring device, can be space-like related, locality is a requirement of relativity theory. (Jarrett 1984, p. 573)

But just as the preparation of one measuring device can be space-like separated from the measurement made by the other, so too can the *measurement itself* on one side be space-like separated from the other. Just as the setting of the right-hand detector is an event which takes place in region M, the passage or absorption of the right-hand photon is an event which takes place in region 2. If space-like separation means that the setting cannot influence the other side, then just as much the *outcome* cannot influence the other side. Bell taught us that forbidding superluminal *causation* means accepting factorization. So if Relativity really forbids superluminal causation we are cooked whether we give up parameter independence or outcome independence.

One final application of the "locality"/"completeness" distinction deserves our attention. Don Howard has argued in several papers that Einstein's picture of physical reality contained two separate conceptual elements, which Howard denominates "separability" and "locality" (see Howard 1985, 1989).

"Locality" in Howard's usage requires that the physical state of a system "is unaffected by events in regions of the universe so removed from the given system that no signal could connect them" (1989, p. 227). "Locality" is thus equivalent to Bell's notion of local causality.[15] Howard describes "separability" as follows:

> This is a fundamental ontological principle governing the individuation of physical systems and their associated states, a principle implicit in many classical physical theories. It asserts that the contents of any two regions of space-time separated by a non-vanishing spatiotemporal interval constitute separable physical systems in the sense that (1) each possesses its own, distinct physical state, and (2) the joint state of the two systems is wholly determined by these separate states. (1989, pp. 226–7)

The separability principle is clearly enunciated by Einstein in the passage cited in the beginning of chapter 1 (Born 1971, pp. 170–1). Bell's account of local causality also presupposed some such principle in that it employs conditionalization on, for example, the physical state v of the region N. This tacitly assumes that any clearly delineated spatiotemporal region *has* an intrinsic physical state. (Bell is perfectly aware of this assumption: that is why he discusses theories of *local* beables.) Definition of the notion of local causality may be complicated by the denial of separable states. This issue will come in for further notice when we discuss the nature of the quantum wave-function.

Howard, however, assimilates "locality" to parameter independence and "separability" to outcome independence, which leads him to maintain that quantum theory is local but non-separable (1989, p. 232). This assimilation cannot be correct. Consider our first crude model of a mechanism for violating Bell's inequality. We assume that the photons have some means of superluminal communication, which we may picture as the tachyon of figure 3.9. Each photon carries the same instructions: if it arrives at a detector without having received a message from its twin then it is to flip a fair coin to decide whether or not to pass. It then communicates to its partner both the setting of its detector and the result of the coin flip. If it has received a message it is to act accordingly: agreeing with its twin if its own polarizer is set at the same angle, agreeing 75 percent of the time if the difference of angles is 30°, agreeing 25 percent of the time if the difference is 60°. This is a completely separable theory: the photons and the tachyon all have perfectly determinate intrinsic states at all times, and the joint physical state of two distinct regions or systems is just the sum of their individual states. Still, it displays a violation of outcome independence and not parameter independence. Conditionalizing on the setting of the distant polarizer does not change the probabilities at the local one, conditionalizing on both the setting and the result does. Furthermore, if we cannot control the stochastic element and cannot observe anything more than the passage or absorption of a photon then we cannot use this mechanism to send signals.

We might note in passing that the objections raised in chapter 3 to using tachyons to send signals disappear in this context. Tachyons caused problems because if used to send messages back in time they would have negative energy and so create dynamical instabilities. But there is no reason the tachyons must transmit energy to send signals. A massless tachyon could be used here. Massless particles are constrained to stay on the light cone only if they are to carry energy, as photons do. Further, only a massless particle can transmit a finite amount of energy at light speed since γ is singular for $v = c$. The energy and momentum of a superluminal massless particle would be zero.

Violations of Bell's inequality show us that the world is not causally local, Jarrett's analysis notwithstanding. Under plausible assumptions about what humans can control and observe, it seems that this violation of local causality cannot be employed in a superluminal signaling device. We have also seen that the prohibition of superluminal signals is often conflated with the prohibition of superluminal causation, and that the constraint on signaling is not infrequently taken to be the fundamental relativistic condition. In particular, the Equal Time Commutation Relations of relativistic quantum field theory amount to no more than a prohibition of signaling faster than light. So we must finally face the question: what would be wrong with a Bell telephone? With what relativistic principle would superluminal signals conflict? This question will bring us closer to the heart of the Theory of Relativity.

Relativity and Signals

The claim that Relativity forbids signals which travel faster than light is often made without justification and accepted without demur. The idea draws strength from the operationalism which infected many early expositions of the theory. The problem of synchronizing distant clocks was set, and the most obvious means for doing so employed signals sent between the clocks. A typical argument would run as follows.[16]

Let A and B be identically constructed clocks (figure 4.2). The problem is to synchronize A with B. A signal is sent from A to B at t_0 whose arrival

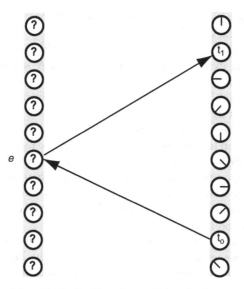

Figure 4.2   Sychronizing Clocks by Signals

at B may be called event *e*. As soon as it arrives a return signal is sent from B to A which arrives at time t_1. If we now employ only the principle that *a signal must arrive at its destination at a time later than it is sent* we can infer that the event *e* must occur after t_0 and before t_1, but no finer discrimination can be made. Any time between t_0 and t_1 can be assigned to *e* without violating the constraint. This indeterminacy can be reduced by employing faster and faster signals. In the limit, if the interval between t_0 and t_1 can be made vanishingly small, the time assigned to *e* will be fixed to arbitrary accuracy. Classical physics, with no limit on signal speeds, supports this univocal method of synchronizing clocks and hence supports an absolute notion of simultaneity. If light is the fastest signal that can be sent, though, the interval between t_0 and t_1 cannot be reduced beyond a certain point. Hence many different assignments of times to *e* will always be allowable and the various choices reflect the relativity of simultaneity in the Theory of Relativity. Or so the story goes.

This result should be surprising to us since the speed of light played the role of an invariant rather than a limit (of signals or anything else) in our derivation of the Lorentz transformation. Further, the argument given above relies on a principle which we never needed to invoke, viz., that the emission of a signal must always come prior in time to its reception. If one accepts this principle along with the equal validity of all Lorentz frames then the restriction on the speed of signals immediately follows, for in some Lorentz frames the reception of a superluminal signal will precede its emission: the signal will travel backwards in time.

The reinterpretation principle which saved tachyons from propagating backward in time can be of no avail for signals. A signal requires a correlation between a controllable physical state and an observable one, the source of the signal being identified not by its position in time but by its controllability. Let us take as the emblem of a signaling mechanism a button which can be pushed and a lamp which goes on and off. If there is a lawful correlation between the button being pushed and the light going on then the mechanism can be used to send signals. The button can be treated as a free variable in the sense that it can be coupled to any manner of device which will determine its state. If the laws of nature imply that the light flashes in the same sequence as the button is pushed, then signals can be sent from the region of the button to that of the light.

When we say that the state of the button can be treated as a free variable we do not mean to conjure up the idea that the button pushers must have free will in some deep metaphysical sense. Indeed, a good experimental proof that a signaling mechanism exists is not provided by allowing a human to push the button at mere whim, for in that case one might wonder whether the button is causing the lamp to light or the lamp lighting is somehow influencing the experimenter to push the button. A much better test is rather provided by coupling the button to a mechanism whose workings are predictable and well understood, by, for example, a computer programmed to output *The Wasteland* in Morse code. If the light flashes out the poem then (assuming common causes have been ruled out) the causal arrow must go from button to light since the state of the button is already adequately accounted for irrespective of the state of the light.

So if superluminal signals exist, then in some Lorentz frames signals can be sent back in time. More generally, in some Lorentz frames causes can precede their effects, since signals are a species of causal process. But why should we suppose this to be either physically or metaphysically objectionable? What grounds do we have which allow us to impose any temporal constraints on signals?

One possibility is to claim that temporal ordering is somehow definitionally connected to the very notion of a signal. This might be because time order is used in defining what a signal is or because time order is itself determined by signals. The first possibility has already been discounted: we have defined signals above without reference to time order. The idea that time order derives from the prior notion of signaling is much more subtle. Hans Reichenbach's general account of time order follows something of this pattern. Reichenbach begins with a definition of a causal process as a process which carries marks. That is, a modification at one end of the process is always associated with a correlative modification at the other, while the converse does not hold. If the modification is controllable and observable, such as a mark on a stone which is thrown from one point to another, then such a causal process can serve as a signal, and indeed Reichenbach uses "causal chain" and "signal chain" interchangeably (1956, pp. 137–8). Causal chains are asymmetric since the mark is only transmitted in one direction, and Reichenbach defines the notions of "earlier" and "later" in terms of this asymmetry. It might seem, therefore, that signals which are received before they are sent are a conceptual impossibility: the reception is by definition later.

Our problem, however, is not so radical as Reichenbach's. We are not concerned to define the notion of temporal order *ab ovo*: we are already starting with a well-entrenched framework of spatio-temporal and causal relations. That framework includes, among other things, local inertial frames with temporal coordinates, causal explanations of the operation of computers, and so on. The introduction into that framework of buttons and lights which behave as described above is no conceptual impossibility, even if the light should happen to flash before the time the button is pushed.

Let us suppose, then, that we have such a device. For every time the button is pushed there is a correlated instance of the light going on. Furthermore, the event of the light going on is always at space-like separation from the pushing of the button, so in some Lorentz frames the signal is sent at a later time than it arrives. Would this constitute a violation of Relativity?

The invariance of the speed of light in all frames would obviously be unaffected. The velocity at which the *signal* is sent would exceed the speed of light in every frame, but would not be invariant. In some frames the signal will be sent instantaneously, in others at a finite superluminal speed, and in yet others, those in which the signal goes back in time, the speed would be a bit tricky to define. But subluminal speeds are also not invariant, so this is no worry.

The important question is whether the phenomenon would allow us to pick out a particular Lorentz frame as holding a privileged position in nature: if so, then a fundamental relativity principle will be violated. But that depends on the details of the superluminal transmission, on the exact connection between the emission of the signal and its reception. The laws of transmission could take more forms that we can usefully survey, but three cases will suffice to illustrate the point.

Case 1

The point of signal reception may be determined by the state of the transmitter (or by the state of some other physical system). For example, the signal could be sent at some fixed velocity (including instantaneously) *in the frame of reference in which the transmitter is at rest*. The case of

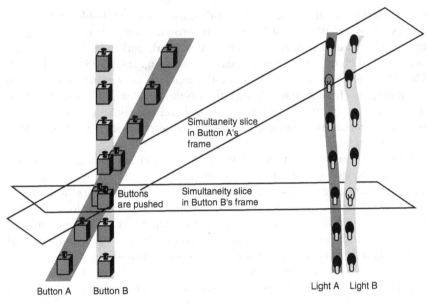

Figure 4.3   Superluminal Signals Dependent on Source

instantaneous transmission is illustrated in figure 4.3, where each button controls the corresponding light. The spatio-temporal careers of the lights have been only vaguely indicated since we suppose that the points where the signal is received is not a function of the light's state of motion. Rules for signal transmission which do depend on the motion of the receiver could be devised, but this simplest case illustrates our point well enough.

This situation retains the relativity principle fully: no Lorentz frame is intrinsically preferred over any other. It is true that the signal propagation may be isotropic only in the rest frame of the emitter, but so too are water waves isotropic only in the rest frame of the water. The laws of nature do not pick out any particular frame, only the distribution of matter does.

Case 2

The point of signal reception may depend deterministically only on the location of the emitter and not on its state of motion. One possibility is depicted in figure 4.4: if buttons A and B are both pressed at some space-time points then their corresponding lights, if kept together, will go off together.

By far the most interesting form of case 2 is when the points where the signal can be received are independent both of the physical state of the emitter *and* of the receiver. Let us further specialize to cases of *signals which can be transmitted through a vacuum*. In such a case, there will be a unique

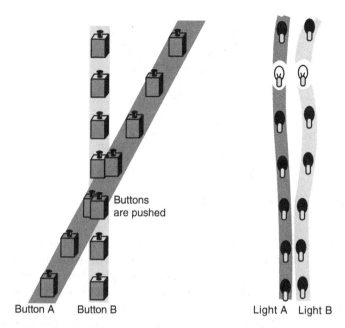

Figure 4.4 Superluminal Signals Independent of Motion of Source

locus of points associated with each point P of space-time, viz., the points at which a signal S emitted from P would be received supposing that the universe contained only an emitter at P and a receiver. Let us call this the *S-signal reception locus* of P.

Since we are imagining signals which are transmitted through a vacuum and unaffected by the state of motion of the source and receiver, the signal reception locus must be a function solely of the point P and the intrinsic structure of space-time. Here we are finally able to make direct contact with Relativity: Special Relativity asserts that the intrinsic structure of space-time is that of a four-dimensional infinite manifold endowed with a Minkowski metric *and nothing more*. So if Relativity is true, the signal reception locus of a case-2 signal would have to be a locus which can be determined given the point P, the metric, *and nothing more*. What are the possibilities for such a locus?

Every hypersurface one could define given only a point and the Minkowski metric as data must be invariant under all Lorentz transformations which leave the point invariant. For the metric is invariant under all Lorentz transformations, and if the data are unchanged by a transformation and the data yield a unique construction then the construction must also be invariant. So which hypersurfaces in Minkowski space-time are invariant under Lorentz

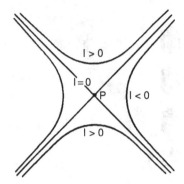

Figure 4.5   Loci Definable From a Point and the Metric

transformations? The different sorts are illustrated in figure 4.5 for a two-dimensional space-time. To get the three-dimensional space-time one rotates the figure about the central axis, forming the light-cones and three hyperboloids of revolution. These surfaces each consist of sets of points with constant invariant interval I from P. The light cone, of course, has I = 0. The loci of constant distance from a point in two-dimensional Euclidean space are circles, the solutions to the equation $\Delta x^2 + \Delta y^2 =$ constant. The loci of constant invariant interval from a point in two-dimensional Minkowski space-time are hyperbolas, solutions to the equation $c^2\Delta t^2 - \Delta x^2 =$ constant. The appearance of the minus sign in the definition of the invariant interval accounts for the fact that hyperboloids are Lorentz invariant surfaces.

Not surprisingly, the light-cones are hypersurfaces invariant under Lorentz transformations. They must appear in our analysis since light is a case-2 signal. The future light cone of the point P is just the light-signal reception locus for P. (The future light cone alone is an invariant surface since we have tacitly restricted ourselves to the *orthochronous* Lorentz transformations, i.e. those that leave the direction of increasing time coordinates unchanged.) The past light cone may be called the light-signal emission locus of P: the set of points from which a light-signal can be sent to P.

The hyperboloid of revolution with I > 0 is a possible signal reception locus in Minkowski space-time. It has several interesting properties. First of all, the signal is always *delayed*: even a receiver which is located next to the transmitter will not get the signal until after the button is pushed. How fast would such a signal travel? Here we must make a careful distinction. When we define the speed of a signal (in a reference frame) we mean the distance from the emitter to the receiver divided by the time elapsed between emission and reception (in that frame). In any given inertial frame the speed of an I > 0 signal is variable: it ranges from 0 upwards, limited by (but never reaching) the speed of light. It is clear that in the most important

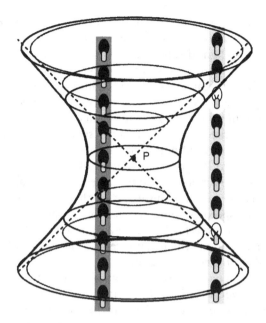

Figure 4.6   A Superluminal Case-2 Signal

sense this signal cannot be superluminal since it can only send messages into the future light cone of P.

There is, however, a way in which the I > 0 signal might seem to be super-luminal. Supposed we packed space-time with lamps which would light upon receiving the signal. Then (in four-dimensional space-time) the phe-nomenology associated with pressing the transmitter button would be this (as described from the rest frame of the lamps). After a delay, the lamp next to the transmitter would flash, followed by an expanding sphere of flash-ing lights. The surface of that sphere would expand *faster* than the speed of light, starting off infinitely fast and slowing down so that it approached (but never reached) the speed of light. Thus the apparent wave-front of the signal would be superluminal, but the wave-front is not itself a signal or any kind of causal process. This distinction between the speed of the signal and the speed of the wave-front is not needed for light since the two always coincide.

Being interested in superluminal signals, we are left with the I < 0 hyper-surface. It is just one hypersurface in a space-time of dimension greater than two. The three-dimensional space-time is shown in figure 4.6, along with two signal receptors. The I < 0 signal is even more interesting than the I > 0 case. The receiver on the left, which comes quite near the emitter when it is pushed at P, never gets the signal at all. The lamp on the right will receive the signal twice, with some interval of time between the flashes. The velocity of the signal is variable in all reference frames, ranging from a velocity just over the speed of light through instantaneousness, and then back down toward the speed of light *for a signal that goes backward in time*. The speed of the signal is always greater than that of light, for the reception points are all space-like separated from the source. The wave-front is always subluminal, with a minimum velocity of zero and limits at light speed.

In a space-time packed with receptors the sequence looks like this (from any inertial frame). From infinite past time a sphere of flashing lights has been collapsing toward P at nearly the speed of light. As it approaches P it slows down. At the very moment that the sphere stops collapsing, at the very center of it, the emitter button is pushed. After that the sphere begins expanding, forever gaining speed as it goes, approaching again the speed of light.

The signal emission locus for $I < 0$ case-2 signals is identical with its signal reception locus: any point which can receive a signal from P can also send one to P using the same type of signal.

So in the case of signals whose propagation is not affected by the state of the emitter or receiver, Relativity, far from ruling out superluminal signals, prescribes a very precise form for them. Light is confined to the light cones because its speed is always invariant. When we relax this invariance requirement we open up the possibility of other sorts of signals whose behavior is dictated solely by the metrical structure of space-time.

Case-2 superluminal signals do have some notably odd features. They do not propagate from the point of emission. Indeed, points sufficiently near the emitter will not receive the signal at all, so there is a spatial discontinuity or "action-at-a-distance" incorporated in these signals. The signals also contain an essential length scale: at spatial lengths shorter than the scale the signal "disappears." At the other end of the spectrum, if we are far away from the source, many times the scale length, then the signal will appear very nearly to propagate like light. The fundamental difference between $I < 0$ signals and light signals is that light can propagate only along the future light cone of P while an $I < 0$ signal propagates both forward in time (nearly) along the future light cone but also backwards in time (nearly) along the past light cone. Or, perhaps less tendentiously, the sphere of signal reception events not only expands out after the button is pushed, it has also already been collapsing in before the button is pushed.

Of course, the fact that Relativity so constrains the form of superluminal case-2 signals means that conflicts with Relativity can arise. But the conflict would arise not merely because the signal propagates faster than light but because it does not do so in the right way. If the Minkowski metric is the total intrinsic structure of space-time then all case-2 superluminal signals must propagate along hyperboloids. In particular, the existence of case-2 signals that propagate along simultaneity slices, or that propagate directly from the source, would contradict Relativity. Relativistic space-time simply does not contain enough structure to pick out a unique simultaneity slice if the only datum given is the point of emission.

Although Relativity rules out every superluminal signal-reception locus except the hyperboloid, we will concentrate our attention almost exclusively on propagation along simultaneity slices (flat space-like hyperplanes). The reason is forward looking: wave collapse in standard non-relativistic quantum theory is modeled as an instantaneous process, and the most natural substitute for instantaneousness in Special Relativity is propagation along flat space-like hypersurfaces slices. Furthermore, wave collapse is supposed to propagate from the source, or, at least, there are no points which are unaffected by wave collapse because they are too close to its epicenter. So if we seek a locus of points which represent all the places whose physical states are changed by a wave collapse due to measurement at P, a simultaneity slice through P would suit the purpose admirably. The problem, of course, is *which* simultaneity slice? If the signal is case 1 then the slice could be picked out by the motion of the source, the most obvious choice being the slice orthogonal to the source's velocity vector. But for a case-2 signal to propagate along a flat hyperplane there must be more structure added to space-time.

Perhaps there is a third possibility. Although a point P determines no unique flat hyperplane in Minkowski space-time, why must the choice be determined at all? Consider this mechanism: when the button is pushed at P a flat space-like hypersurface is chosen at random and the signal propagates along it. Let us make this the third official genus in our taxonomy.

Case 3

The signal reception locus depends only on the position of the emitter and not its state of motion, but the dependence is stochastic. In this case the signal reception locus (in a particular case) need not be Lorentz invariant, that is, the locus itself could be used to single out a preferred reference frame. A flat space-like hypersurface is the example of a non-invariant locus *par excellence*: it picks out the reference frame in which it is a simultaneity slice. But the argument below will hold for any signal reception locus which is not Lorentz invariant.

Even though a case-3 reception locus need not be Lorentz invariant, Lorentz invariance is still a factor. For the *probability structure* of the stochastic process must be Lorentz invariant if Relativity is to hold. Given only the metric and a point P, a case-3 signal mechanism would fix the probabilities for various reception loci to be chosen. The argument for this is the same as that in case 2. The stochastic process by which the probability for a locus to be chosen is determined must not itself single out a preferred reference frame.

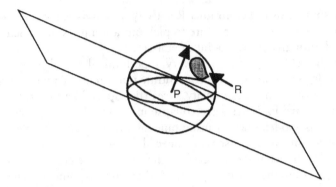

Figure 4.7   Randomly Choosing a Plane in Euclidean Space

Can there be a Lorentz invariant random method for assigning probabilities to sets of flat space-like hypersurfaces through a point P whose only input is the location of P? There cannot.

Intuitions about this case can be easily misled by reading Euclidean structure into our space-time diagrams. In order to forewarn against this pitfall, and to illustrate what invariance requires, let's consider the Euclidean case first. The corresponding question for Euclidean three-space would be: can a method for assigning probabilities to planes through a point P be invariant under rotations about P? Since to every direction in space there corresponds a unique plane, viz., the plane whose normal lies in that direction, our question is equivalent to: is there an invariant method for randomly choosing a direction in space? And since to every direction in space there corresponds a unique point on the unit sphere about P, the question becomes: is there a probability distribution over the points on a unit sphere which is invariant under all rotations of the sphere?

The answer to our last question is trivial: a uniform distribution, and only a uniform distribution, is invariant under all rotations. A uniform distribution is one where for every region R of the sphere, the probability that the chosen point lies within R is proportional to its area (see figure 4.7). We can imagine the actual process of choosing a point as a limiting case of repeated binary choices: divide the sphere into two equal areas and flip a fair coin to choose one. Divide the chosen region in half and repeat the process, and so on ad infinitum. After selecting a large number of points by such a random method it is overwhelmingly likely that the collection of points will not tend to single out any direction in space: the ball will be covered with a uniform density of points. In particular, it becomes overwhelmingly likely that any plane through P will divide the collection of points nearly in half.

Let's now try to apply the same sort of reasoning to our original problem. The task of randomly choosing a flat space-like hyperplane through P is equivalent to randomly choosing a unit vector inside the future light cone of P,

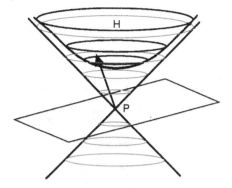

Figure 4.8 Choosing a Hyperplane in Space-time

namely the vector orthogonal to the surface. And that is tantamount to find-ing a probability measure over the points on the hyperboloid H (figure 4.8).

But there is no probability distribution over time-like directions or over points on the hyperbola which is invariant under all Lorentz transforma-tions. Imagine the hyperbola H bisected by the y′–t′ plane (in some inertial coordinate system). By symmetry, the probability that the point chosen will fall on a given side of the plane must be 0.5, otherwise the distribution will already favor some directions over others. But since the distribution is Lorentz invariant, the same must hold for the y′–t′ plane, where we have used the Lorentz transformation (figure 4.9). If there is a 0.5 probability that the chosen point will right of the y–t plane and a 0.5 probability that it lies to the right of the y–t plane, then there is zero probability that it lies between the two planes. Since the entire hyperbola can be covered by a finite number of such regions of zero probability, the probability that the point will fall on the hyperbola at all is zero.

In fact, the hyperbola can be covered by just two such regions. If we think of the vector as a velocity vector, the region in figure 4.9 between the planes includes all the vectors representing a particle whose x-velocity in the unprimed frame lies between 0 and some maximum determined by the tilt of the x′–t′ plane. By tilting the right-hand plane further to the right and the left-hand plane to the left (using Lorentz transformations) we can expand the forbidden region to include all vectors whose x-velocity in the unprimed frame lies between, say, –0.9c and 0.9c. By an analogous argu-ment we can forbid all vectors whose y-velocity lies between –0.9c and 0.9c. But no vector in the future light cone lies outside both these regions: to do so it would have to represent a particle which, in the unprimed frames, is traveling at more than 90 percent the speed of light in both the x and y directions. Its total velocity in the unprimed frame would therefore exceed the speed of light, and the vector would not lie within the light cone.

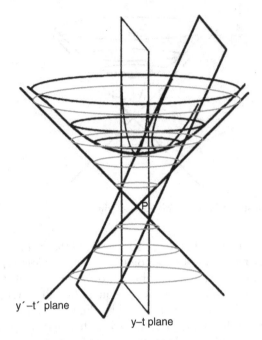

y´–t´ plane

y–t plane

Figure 4.9   Two Planes Bisecting the Hyperbola

In sum, case 3 yields the same result as case 2. If there is a signal whose signal reception locus (in a vacuum) depends, either deterministically or stochastically, only on the position of the emitter and not on its state of motion, then the locus must be a hyperboloid in space-time if the only intrinsic structure of space-time is the Minkowski metric. Put the other way round, the existence of such a signal whose signal reception locus is not a hyperboloid would force us to add more structure to space-time, structure which could be used to single out a preferred reference frame. In this case the relativity principle would be false.

On the other hand, Relativity *per se* in no way constrains case-1 signals. It is consistent even with case-1 superluminal signals which propagate instantaneously (i.e. along a flat space-like hyperplane). But the hyperplane must be in part determined by the state of the emitter, or of some other matter.

Signaling Paradoxes

So far we have examined a rather narrow question: is there anything about superluminal signaling *per se* which is incompatible with the relativistic account of space-time structure? In particular, we asked whether superluminal signals would necessarily imply a preferred inertial frame, and hence a preferred (or even absolute) notion of simultaneity. If this had turned out to be the case then superluminal signals would indeed have been prohibited by Relativity.

There is, however, an entirely different line of reasoning against faster-than-light signaling. Superluminal signals must (in at least some reference frames) propagate into the past: the signal is received before it is sent (as judged from those frames). This in turn is supposed to allow paradoxical situations: I can decide to send the signal if and only if I haven't already received it, leading to a set of unsatisfiable conditions.

The conditions required for the possibility of such paradoxes are more complex than merely the existence of superluminal signals. This is perhaps most easily seen by returning to the superluminal case-2 signal of figure 4.6. As we noted, the flashes of light which indicate the reception of the signal form a sphere which has, for all time, been collapsing in toward the emission event. Why couldn't the signaler adopt the strategy of refusing to push the button if he sees the collapsing sphere? The answer in this case is that he won't see the sphere until too late: none of the light from the lamps will reach the emitter until after the button has been pushed. This is obvious since each of the reception points is space-like separated from P: light cannot travel between the two. A world which contains but one superluminal case-2 signal will engender no paradoxes even though the signal is judged as traveling back in time, not only in some reference frames but in all reference frames.

The problem, then, is not signaling into the past *per se*, the problem is signal loops. If a signal sent from P can also be received at P then the emission can be made dependent on the reception in apparently paradoxical ways. In order to explore the question of signal loops we will need to expand our technical machinery a bit.

Let us call the locus of all points in space-time which can be reached from P by either a single signal or a chain of signals the *domain of signal reception* from P. If B is in the signal reception locus of A and C is in the signal reception locus of B then a signal can be sent in two steps from A to C: we need only have the emission of a signal from B depend on the reception of one from A. If light were the only existing means of signaling then the domain of signal reception of every point in space-time would be the future light cone of the point together with the interior of the cone. (Signals could be sent to the interior of the cone by reflection or retransmission from a point on the surface of the cone.) When referring to domains of signal reception or transmission we will take "light cone" to mean "light cone plus interior."

If only one superluminal signal is allowed in the history of the world depicted in figure 4.6 then the domain of signal dependence of P will be the union of the future light cone of P with the future light cones of all the points on P's signal reception locus. That union constitutes the complement

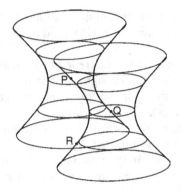

Figure 4.10   Signal Chain into the Past Light Cone

of P's past light cone. Since the domain of signal emission to P is exactly its past light cone, this means that for any point in space-time either a signal can be sent to or received from P. But no point can both send and receive such a signal.

If superluminal signals can only be sent from P then the domain of signal dependence for any point outside P's past light cone is just its own future light cone. The domain of signal dependence of any point in P's past light cone is the union of its own past light cone with P's domain. In no case is a point in the space-time within its own domain of signal dependence, so no signaling paradoxes can arise.

The situation changes radically if the number of superluminal case-2 signals that can be sent is unlimited and if the signals can be sent from anywhere. Then a signal can be sent from P into its own past light cone in two stages: first from P to a space-like separated point Q and then from Q to R (figure 4.10). The domain of signal reception of P, as of every point in the space, is the whole space. In particular, the domain of P includes P itself. A signal can be sent from P to Q to R and then, by a subluminal messenger, back to P.[17] It therefore seems that we can construct a mechanism with paradoxical consequences. A signal will be sent from Q if and only if one is received from P. A signal will be sent from R if and only if one is received from Q. And a signal will be sent from P if and only if one is *not* received from R. It seems that the signal neither can nor cannot be sent from P.

The analysis of such loop paradoxes is reserved for the next chapter. At this point we only want to establish that the paradox arises only if a point lies in its own domain of signal reception, and that this condition is not equivalent simply to the existence of superluminal or backwards-directed signals. However, we have seen that Relativity does imply that all case-2 signals must have hyperboloids as their domain of signal reception.

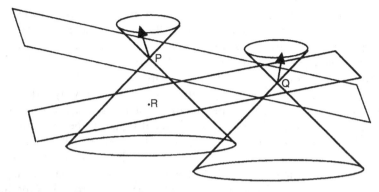

Figure 4.11   Case-1 Signal Loop

Hence *Relativity implies that if case-2 superluminal signals can be sent from any point in space-time, then signal loops are possible.*

What of case-1 signals? In particular, what of case-1 signals whose signal reception loci are flat space-like hyperplanes? Since we are assuming that the state of the emitter determines the hyperplane of transmission, signals can propagate along planes which are tilted with respect to one another. Indeed, if the rule stating the dependence of the hyperplane on the state of the emitter is itself to be Lorentz invariant (and hence does not implicitly single out a preferred reference frame), any hyperplane must be able to serve as a plane of propagation. But then by controlling the state of the emitters we can construct the situation depicted in figure 4.11. We have assumed for simplicity that the signal reception locus is the hyperplane orthogonal to the velocity vector of the emitter, but the same result occurs in any relativistic theory with case-1 superluminal signals. Again, a signal can be sent from P to Q to R and back to P, closing the loop. And again the domain of signal reception of every point in the space is the whole space.

In order to have superluminal signals whose locus of signal reception is a flat hyperplane and yet to avoid the possibility of loops the hyperplanes must stack up, like the simultaneity slices in a particular frame of reference.[18] In fact, the loci would *be* simultaneity slices in a particular frame, so such a theory is manifestly non-relativistic. The case is not helped if one switches to something other than flat hyperplanes. One could, for example, use cones which are flatter than the light cone but not entirely flat. But once again, such loci are not Lorentz invariant (as the light cone is) so they must be determined by something other than the metric. If the extra is the state of the emitter then by tilting the cones one can get loops. And if it is rather some global constraint, such as requiring that the cones all stack up without tilting, then we must import extra spatio-temporal structure, contradicting Relativity.

So the choice we have is not between superluminal signals and Relativity but between superluminal signals which allow loops and Relativity. If we can argue that loops are impossible, then accepting the signals means positing extra structure to space-time to forbid them, structure that looks suspiciously like an absolute notion of simultaneity.

This may seem like a very long digression on an empty hypothesis. We began by demonstrating that the violations of Bell's inequality predicted by quantum theory do not allow superluminal signals to be sent. But all of the work we have done is not in vain. We also examined Bell's argument that the inequalities cannot be violated by any theory in which all of the causes of an event lie in its past light cone. So even though nature may not allow superluminal signaling, it does employ, according to Bell, superluminal causation. And if the analysis just given can be repeated verbatim for superluminal causes we are going to be left with a choice between violating Relativity or countenancing *causal* loops. If causal paradox forecloses accepting such loops, or if they are objectionable on other grounds, then the observed violations of Bell's inequality will pressure us to add more structure to space-time than Relativity says there is.

Appendix B: Bohmian Mechanics

In this chapter we have used David Bohm's interpretation of quantum mechanics as an illustrative example of a theory which yields predictions of the puzzling quantum correlations. Since Bohm's theory is not so widely known, and is often misunderstood, it seems apposite to present a brief account of the main points of his approach. As usual, we will minimize the technical details in favor of the fundamental ontological picture.

It is easiest to present the basic tenets of Bohm's theory by contrasting it with orthodox quantum mechanics, as typified by von Neumann's classic presentation (1955). Orthodox non-relativistic quantum theory employs three leading principles:

1 The wave-function is *complete*, i.e., the physical state of a system is entirely captured by the wave-function. Further, a system only possesses an observable property if the wave-function of the system is an eigenstate of the associated Hermitian operator. Corollary: two systems described by the same wave-function are physically identical.

2 The wave-function usually develops in accord with Schrödinger's equation, but does not always do so. Sometimes the wave-function "collapses." (Orthodox quantum theory has no generally accepted account of how or when this collapse occurs.)

3 The complete dynamics is stochastic rather than deterministic. Isolated systems which begin in identical states may end up in different states.

The stochastic element of orthodox quantum theory appears in the wave collapse since Schrödinger evolution is deterministic.

The leading ontological principles of Bohm's theory are simply the negations of 1, 2 and 3:

1 The wave-function is not complete. Apart from the physical factors reflected in the wave-function, a system has other physical properties. In Bohmian particle mechanics, particles always have definite positions; in Bohmian field theory, fields always have definite values. It follows that systems sometimes have definite observable properties even when not in eigenstates of the associated operators, e.g. a particle can have a position even its wave-function is not an eigenstate of the position operator. Further, two systems can have identical wave-functions but be in different physical states since the values of these other properties may differ. These extra physical parameters which are not reflected in the wave-function are commonly (and misleadingly) called "hidden variables."

2 The wave function always develops in accord with Schrödinger's equation. There is no wave collapse.

3 The dynamics is deterministic. Since the wave-function is always governed by Schrödinger's equation, this is obvious for that part of the dynamics. In Bohm's theory, the "hidden" parameters also evolve deterministically.[19]

The basic ontology of a Bohmian theory is very clear and straightforward. In Bohmian particle mechanics, for example, there are particles. These particles are rather like classical particles in that they always have definite positions and trajectories. Tables, chairs, cats, and people are made up of collections of particles in table-, chair-, cat-, and person-shaped configurations. The Schrödinger's cat problem cannot arise for Bohmian cats: it is generally quite simple to distinguish configurations of particles which constitute a live cat from those that constitute a dead cat. This is so even though the *wave-function* for the cat may be a superposition of a part corresponding to a live cat and a part corresponding to a dead cat.

In addition to the particles, Bohm's theory recognizes a real physical object which is described by the wave-function. This object has some peculiar properties, but for the moment let us focus on its positive characteristics. This object, which we may call the wave-function of the system (eliding the distinction between the object and its mathematical representation, as is commonly done), is not made up of particles. It always evolves in accord with Schrödinger's equation. And, most importantly, it determines how the particles will move. The mathematical heart of Bohm's theory, then, consists in two equations. The first, Schrödinger's equation, determines how the wave-function changes. The second, which we give below, specifies how the particle positions will change given the wave-function. We specify the physical state of system by specifying both the wave-function of the system and the positions of its particles, and the two equations then determine how the system will evolve. End of story.

Explicitly, the two fundamental equations are Schrödinger's equation:

$$ih\frac{\partial\Psi}{\partial t} = -\frac{h^2}{2}\Delta\Psi + V\Psi$$

and Bohm's equation:

$$\frac{dq}{dt} = h\,\mathrm{Im}\frac{\mathrm{grad}\psi}{\psi}(q),$$

where $q = (q_1, ..., q_N)$ specifies the position of particles 1 through N, Ψ is the wave-function, and V is the potential energy of the system. As Bohm's equation shows, the wave-function directly determines the velocities of the particles. Bohm's equation is rather easily motivated if one assumes that the wave-function should determine such a velocity field. Indeed, given such desiderata as Galilean, time-reversal, and rotation invariance, and that wave-functions which are proportional should be physically equivalent, the derivation of the equation is almost inescapable (see Durr, Goldstein, and Zanghì 1992, pp. 851–6).

None of the fundamental interpretive problems of orthodox quantum theory exists for Bohmian mechanics. There is no "measurement problem." Measurements are physical interactions like any others; measuring devices, just collections of particles. Depending on the state of the system being measured the measuring device will end up in one or another distinct state. There is no wave collapse and so no problem about when or how it occurs. But several obvious questions must still be addressed. How could Bohmian mechanics be in any way observationally equivalent to quantum theory, given that quantum theory is stochastic and Bohm's theory deterministic? Where do probabilities come from in Bohm's picture? How can Bohm's theory give the same predictions as quantum theory if there is no wave collapse in the former but there is in the latter? And lastly, if Bohm's theory is devoid of interpretational difficulties, why has it not simply displaced orthodox quantum theory?

The first questions are easily answered. A deterministic theory yields definite non-probabilistic predictions only given the precise initial state of a system. If one is unsure about the precise initial state, then that uncertainty will infect predictions about the evolution of the system. A probability function defined over the possible initial states will be deterministically evolved into a probability function over possible final states.

Bohm's theory presupposes an uncertainty about the exact initial positions of the particles in the system. Indeed, Bohm assumes that the original uncertainty is proportional to $|\psi|^2$, the square of the wave-function. This assumption is *not* due to any fundamental connection between the wave-function and probabilities, rather it must be motivated in some other way. Bohm sometimes supposes that the justification may arise from some sub-quantum level of truly random fluctuation (see Bohm 1957, pp. 112ff.). Durr, Goldstein, and Zanghì (1992) argue that the choice of $|\psi|^2$ as the initial probability distribution can be justified without any stochastic processes by a close consideration of the notion of universal typicality. In any case, the important result is that *if the probability function $|\psi(0)|^2$ is used for the initial configuration then the probability for finding a configuration will be $|\psi(t)|^2$ at any later time t.* These are exactly the probabilistic predictions given by orthodox quantum theory, so Bohmian mechanics is, under these circumstances, empirically equivalent to orthodox quantum theory. Further, the Bohmian dynamics precludes determination of the particle positions to a degree of precision which would allow more exact predictions than those available in the orthodox theory.

The status of wave collapse in Bohm's theory is also easily explained. The wave-function is defined over configuration space, that is, over a space each single point of which specifies an exact configuration of all of the particles in the system. (This is also true in orthodox quantum theory, a fact often overlooked because, for a single particle, configuration space is isomorphic to physical space.) The motions of the particles in the system are all codified in the motion of a single point in the configuration space, so we can regard the history of the particles as a single trajectory through this space. Now the important point in Bohm's theory is that the dynamics are, as it were, local *in configuration space*. That is, the trajectory of that point in configuration space is determined by the value of the wave-function *around that point*. Portions of the wave-function in other regions of configuration space have no immediate influence on how the particle positions evolve.

After a Schrödinger's cat experiment, the wave-function for the cat has two large pieces: one in a region of configuration space corresponding to a live cat and one in a region corresponding to a dead cat. The particles of the cat are either in the configuration of a live cat or that of a dead cat. If they are in the configuration of a live cat (i.e. if the cat is alive) then the piece of the wave-function in the dead-cat region will have no immediate effect on what the cat does. For all practical purposes, that portion of the wave-function can be thrown away, the wave-function "collapsed" to just the bit in the live-cat region. Thus wave collapse is a practical expedient that will generally work even though the wave-function never actually collapses.

The caveat that must be added to this rule of thumb is that sometimes, due to Schrödinger evolution, a seemingly irrelevant part of the wave-function will later contribute to the wave-function in the relevant region of configuration space. This happens, for example, in the classic two-slit experiment: the bit of the wave-function which "goes through" the slit which the particle avoids can later influence the trajectory of the particle. The particle itself only goes through one slit, but the parts of the wave-function associated with the both slits come to overlap in configuration space, resulting in the famous interference fringes.

Finally, why has Bohm's theory not been universally adopted? This is a tortured question, but deserving of some notice. Two main complaints have been leveled against the theory: first that it is *ad hoc*, and second that it is non-local.[20]

The charge of being *ad hoc* is a bit hard to evaluate, being rather subjective. Bohm's equation is very simple and quite naturally derived. Bohm's first papers (1952a,b), as well as some of his later work, employed an unnecessary device which was regarded as artificial. This device was the so-called *quantum potential*, a quantity which is definable from the wave-function. The quantum potential was supposed to influence the particles, guiding them in their trajectories. The potential was sometimes thought to be "cooked up" just to replicate the predictions of quantum theory.

This charge is unfounded on two counts. First, Bohm's theory is quite immediately and naturally suggested by formal analogies between the quantum equations and equations found in classical Hamilton-Jacoby theory. Quantum equations yield conserved currents which look formally just like classical probability currents, and immediately suggest a velocity field on configuration space. Interpreting these equations in light of classical Hamilton-Jacoby theory suggests the existence of a non-classical potential, the quantum potential. Furthermore, thought of in this way, Bohm's theory provides a clean criterion for when quantum theory will diverge from classical mechanics: namely when the quantum potential is not negligible.[21]

But the deeper defense against criticisms of the quantum potential is that it is superfluous. In Bohmian mechanics the wave-function determines the evolution of the particles via Bohm's equation: no further potential is needed. The temptation to introduce the potential only arises *if one wants to recover Newtonian dynamics* (as is built into Hamilton-Jacoby theory). In the Newtonian milieu, changes in particle velocities must be due to forces, and the quantum potential provides just the forces that the particle trajectories require. But a radical Bohmian mechanics can just reject Newtonian dynamics altogether. Particles are guided by the wave-function, not pushed around by forces.

Indeed, in this light Bohm's theory can be viewed as a radical rejection of the Newtonian picture. For Newton, the fundamental variables which fix the state of a system are the *positions* and the *velocities* of the particles. Potentials then allow one to derive the accelerations of the particles (via forces), and these elements yield determinate trajectories. The velocity is critical: a "snap-shot" of the world which reveals the positions of particles and the forces at work will not fix their trajectories. But in Bohm's picture the fundamental variables are rather the positions of the particles and the *wave-function*. Velocities are dynamically meaningless: a snapshot of the world which fixes positions of particles and the wave-function at a time determines the entire future evolution of the system. The potentials determine how the wave-function evolves and the wave-function determines how the particles move. Constructing a quantum potential designed to reintroduce Newtonian dynamics is, from this perspective, pure anachronism.[22]

As to the charge of non-locality, Bohm must plead guilty. Given the clarity of Bohm's picture, and the fact that the theory is deterministic, it is rather easy to show how an event in one place can have immediate effects at arbitrary distance. But the burden of our treatise is to show that non-locality is unavoidable in *any* theory which recovers the predictions of quantum theory. Violations of Bell's inequality show that the *world* is non-local. It can be no criticism of a theory that it displays this feature of the world in an obvious way.

Notes

1   The "could" is notable here. The argument for dynamical instability clearly depends on the existence of a coupling interaction between normal matter and tachyons which allows the former to emit the latter. If no such coupling exists, no instability would arise. Of course, if no interaction between tachyons and normal matter existed at all, tachyons could not be invoked to explain any of the observed behavior of our measuring instruments. It would be worthwhile to explore the range of interactions or couplings which would be non-trivial but also not catastrophic.

2   The situation here displays a striking analogy to the interpretational problems surrounding the quantum theory. As Bell has insisted, no acceptable physical theory can be formulated using such terms as "system," "apparatus," "environment," "microscopic," "macroscopic," "reversible," "irreversible," "observable," "information," or "measurement," as all of these are imprecise (Bell 1990, p. 19).

3   This section presupposes some familiarity with quantum formalism. Readers mystified by it can just skip to the conclusions.

4   This is the so-called Schrödinger picture of quantum formalism. In other pictures either part or all of the time evolution is assigned to operators rather than to the quantum state.

5   The commutation relations for the quantum field can be derived from some rather complex considerations of the transformation properties of quantum states under the Lorentz transformations, including time reversal and parity. See, for example, Schweber 1962, pp. 177–80.

6 The qualifier "spatio-temporally separated" is added here to avoid a quibble raised by Bas van Fraassen. van Fraassen argues (1980, pp. 18ff.) that Reichenbach's principle of causal explanation is incompatible with any fundamentally stochastic theory since such a theory will exhibit correlations which cannot be explained by any Reichenbachian common cause. His example involves a stochastic transition to one of a set of states $\{F_1 \ldots F_n\}$, where a state characterized by F_i is also always characterized by some other attribute G_i. If, given some initial state S, the probability that the state F_i will occur is $1/n_i$, then the probability that G_i will occur is also $1/n_i$. But the probability that the joint state F_i and G_i will occur is not the product of the probabilities of F_i and G_i separately: prob $(F_iG_i/S) = 1/n_i \neq$ prob$(F_i/S) \times$ prob(G_i/S). van Fraassen seems to have in mind a case where the coming into existence of the state F_i spatio-temporally coincides with the coming into existence of the state G_i; otherwise the one which comes first can be taken to be the (deterministic) cause of the other and the factorization condition will hold. The mode of mutual determination of spatio-temporally coincident events (i.e. whether one can "cause" the other) is a somewhat puzzling question, but one which has no bearing on our problem, van Fraassen's objection has force only in such cases.

7 As discussed in chapter 5, this "principle" is not really a demand for a particular type of "causal" explanation, but rather a consequence of the notion of a cause. That is, such reliable correlations will validate certain counterfactual claims, and those counterfactual claims imply that there is some causal connection.

8 Determinism does follow if we now suppose the strict correlation or anticorrelation of results when measurements are made in the same directions.

9 Characterizations of locality for stochastic theories are surveyed by Geoffrey Hellman (1982). Hellman arrives at a rather more complicated definition than ours, but ours covers all of his cases. Some of his cases are covered by the fact that we conditionalize on the physical state of the *entire* past light cone of the region of the event (cf. Hellman's condition 11'). Hellman presents a supposed counterexample to this analysis which relies on a "particle" which follows a stochastic discontinuous trajectory. It is clear, however, that such behavior *can* be accounted for by a locally causal theory: simply have the stochastic choice about where the particle will appear next made at the previous point in its trajectory. Since Bell's theorem discusses the *possibility* of accounting for behavior by a locally causal theory this example presents no problem. Indeed, the behavior described can be recovered by a deterministic theory. If Hellman insists that his interest is in *one particular* theory, a theory according which the state of the particle at point n of its trajectory does *not* determine its next appearance, then the appropriate response is that such a theory *is not* locally causal.

10 k can be stipulated to include all causally relevant physical facts in the overlap of the back light cone, but the settings of the detectors cannot be *stipulated* to be the only causally relevant physical events in the remainders of the cones. Bell deals with any other factors by averaging over them: the reader is referred to his treatment (1987, p. 57).

11 Shimony credits this proof to David Mermin.

12 Things are slightly more complicated than this. In Bohm's theory there is a preferred reference frame, and the results of measurements are determined not only

by the state of the particles and the settings of the detectors, but by which detection event occurs first in the preferred frame (we will take up this issue in detail in chapter 7). Still, given all this information, outcome independence holds since to be told the outcome is to be given redundant information.

13 Jarrett's conditions differ a bit from Shimony's since Jarrett employs probabilities conditional on the information that no distant measurement is made rather than probabilities which contain no information at all about the distant setting or result. This does not affect our analysis.

14 Jarrett appends the following footnote to his argument: "While it is here assumed that the experimenters are able to prepare systems in the desired states, the inability to do so in practice does not undermine the argument being made" (1984, p. 587). No support is offered for this claim.

15 Howard does not see his stated locality condition as equivalent to Bell's, since he thinks that "locality" implies only outcome independence and not factorizability. But the stated condition is clearly the same as the one Bell uses to derive factorizability. Howard has been misled by an attempt to map his distinction on to Jarrett's, as we will see below.

16 This form of argument is found, for example, in Reichenbach 1956, pp. 143ff.

17 There is also the simpler "loop" which consists of sending a signal from P to Q and one from Q to P. But on the plausible assumption that some time must elapse between receiving a signal and taking an action which depends on what is received, the two-step loop cannot be used to generate paradoxes.

18 This is not strictly true, but the counterexamples are so contrived as not to be worth the time to go through them. More importantly, the counterexamples are not Lorentz invariant: they must postulate non-invariant spatio-temporal structure, and so are not relativistic.

19 Theories very like Bohm's in which the extra variables evolve stochastically also exist. Indeed, one can define an infinite class of theories determined by a parameter which quantifies the nature of the stochastic element. See Goldstein 1987.

20 Bohm's theory has some other peculiarities to it, but it is hard to argue that these constitute good reason for ignoring it. It does, for example, take the wave-function very seriously even though it is defined over configuration space, not physical space. But then, the wave-function in regular quantum mechanics is also defined over configuration space. Also, particles in their lowest energy states (for example, electrons in unexcited states) are at rest according to Bohm's theory. Finally, Bohm's theory may have been neglected by physicists who thought that a proof by von Neumann showed hidden-variable theories to be impossible.

21 I owe these observations to Robert Weingard.

22 It is perhaps interesting to note that Bohm's theory is deeply congenial to an ontology which maintains that all which exists is that which exists *now*, i.e. at a point in time classically conceived. Instantaneous velocities can of course be defined, but only as the limit of average velocities over finite periods of time. Those puzzled about the status of velocities in an ontology in which only an instant of time *exists* can happily adopt a Bohmian ontology of particles (with positions) and the wave-function.

5

Causation

In the previous two chapters we have seen that violations of Bell's inequality do not *per se* imply the possibility of sending either energy or signals from one wing of the experiment to the other. Furthermore, quantum theory (along with the usual presuppositions about what can be observed and controlled) entails that energy and signals cannot be sent via the mechanism which produces the violation. On the other hand, outcomes on one side are not statistically independent of those on the other and, as Bell showed, this dependency cannot be accounted for by common causes which lie in the past light cones of the measurement events. As we saw in chapter 1, no strategy for answering questions which would reliably produce the quantum correlations can be devised so long as the answer at each side is produced in ignorance of the question asked of one's partner (i.e. in ignorance of the setting of the distant apparatus). Since conditions can be arranged so that neither the act of setting the distant device nor any other events which would allow us to predict that setting lie in the past light cone of the measurement, the result on one side must depend on space-like separated events.

How should this dependency be understood? In particular, do violations of the inequality demand some direct *causal* link between the two sides of the apparatus? If so, do superluminal causal influences run afoul of Relativity?

Any easy approach to these questions is immediately blocked by the problematic nature of the central concept. Causation has been a subject of intense philosophical scrutiny at least since Hume's skeptical attack on the notion of necessary connections in nature. Any general analysis of causation must face a battery of difficult cases involving causal overdetermination, the role of background conditions, distinguishing causes from effects, correlations

Quantum Non-Locality and Relativity: Metaphysical Intimations of Modern Physics,
Third Edition. Tim Maudlin.
© 2011 Tim Maudlin. Published 2011 by Blackwell Publishing Ltd.

due to common causes, and so on. As we will see, these problems in turn demand consideration of the semantics of counterfactual assertions, thereby unleashing yet another swarm of puzzles.

Fortunately, we need not produce a completely general theory either of causation or of counterfactuals. The case before us involves the most uncontroversial application of these notions, examples on which all acceptable theories must agree in order to be acceptable. We will therefore attend closely to our task, developing only so much theory as is needed. Extending the account of causation to cover more recondite cases would be an unnecessary distraction.[1] Let's begin by surveying the terrain we must explore.

Causation, Counterfactuals, and Laws

The usual preliminary indication of a causal connection between types of events is a correlation between them. But it is also a truism that correlation is not *per se* any proof of a direct causal connection. Correlations may be due to common causes, or they may be entirely accidental. Since accidental correlations may exist without any causal connection at all, direct or indirect, they provide an ideal opportunity to discover what beyond mere correlation is required for causation.

Consider a world which contains some irreducibly stochastic events. For simplicity, suppose that some process, such as flipping a coin, could have either of two outcomes and that the process is fundamentally indeterministic. No external conditions determine which result will occur, nor, we can imagine, do they even influence the likelihood of one outcome over the other.

Such a world is manifestly possible. Or at least, the description above involves no conceptual confusions or contradiction. (According to the usual interpretation of quantum mechanics, our world is such a world: take radioactive decays as the random events.) Now suppose we choose two coins and start flipping them. Even in the absence of any causal connection between them, it is possible that the results of the flips will be perfectly correlated. Indeed, the probability of such a result is calculable. If the process is indifferent between the two results then the probability of perfect match over a series of N flips is $(1/2)^N$.[2]

It is possible, then, to have perfect correlation between instances of event types (e.g. coin A coming up heads and coin B coming up heads) absent any causal connection between them. In the case imagined, causal connections were postulated not to exist and the possibility of correlations derived in accordance with the postulates. Hence correlation does not imply causation. The same could occur in a deterministic world: coins flipped in distant locations could match on account of coincidences in the initial conditions rather than through any causal connection.

We are here concerned not with epistemic but with conceptual matters. No doubt the inhabitants of such worlds, if they should notice the coincidence of the results, would suspect a causal connection. They would be rational to do so. Indeed, if upon further inspection no causal mechanism could be found to explain the persistent correlation, they might well posit some heretofore unknown physical mechanism to create the supposed causal influence between the coins. They would, despite their best efforts, have been deceived by a run of bad luck, a cosmic coincidence. Excellent reasons to believe in a causal connection may exist in the absence of one. Our problem is not yet how, or whether, we can have infallible access to the causal structure of the world. For now we simply want to investigate what makes for a causal connection. Apparently, correlation alone does not suffice.

What, then, distinguishes an accidental correlation between the coins from a causal connection? The missing ingredient must go beyond the statistics of actual results. The obvious, and I think correct, diagnosis of the case of accidental correlation is that although the coins always match, they need not have. The result was accidental rather than due to a causal connection because the laws governing the coins did not imply that any correlation would exist. Or, put another way, in this case the correlation does not support counterfactual inferences. If one coin's result *caused* the other then *had the first been different so would have the second*. Given the description of the case this counterfactual assertion is not true.[3] One key way that causation goes beyond mere correlation, then, is that causal connections support counterfactuals while *de facto* correlations may not.

Now it is one thing to note a systematic and important connection between causation and counterfactuals and quite another to analyze causation in terms of counterfactuals. The simplest attempts in this direction are bound to fail. In cases of causal overdetermination, for example, the usual counterfactuals do not hold. The electrical activity in the primary navigational computer was among the causes of the shuttle's perfect landing, for example, but even had that activity not occurred the shuttle would still have landed in the same way: the back-up computer would have kicked in. (Those interested in the complexities involved in a counterfactual analysis of causation can do no better than David Lewis' "Causation" (1986, pp. 159–213).) Fortunately, we need not attempt a full analysis here. It suits our purposes to discover an uncontroversial *sufficient* condition for causal connection, and we can narrow our focus accordingly.

A passable first approximation is this: given two events A and B, if B would not have occurred had A not occurred (or if B would have been different had A been different) then A and B are *causally implicated with each other*.

This can only be a first approximation since there are counterfactual dependencies which do not involve causation among events. For example, the relational properties of one object may depend on the state of other objects in a way that supports counterfactuals without there being any physical causation between the objects. When Oedipus killed Laius at the crossroads, Jocasta, in Thebes, became a widow. If Oedipus had not killed his father she would not have become a widow. Still, this is not a case of one localized physical event causing another since Jocasta's being a widow is not a local physical property of her. We do not want to waste our time on such *scheinprobleme* as exactly when Jocasta became a widow (instantaneously in some preferred reference frame? when she intersects the future light cone of the murder?), whether there *is* an event of her becoming a widow, and so on. If this is superluminal causation, it is not the sort to be of any concern.

If, on the other hand, becoming a widow were a *local physical* event, or if it had local physical consequences, the problem would become acute.[4] Suppose women became markedly paler (or markedly more cheerful) at the moment of their widowhood. Then we could begin to ask sensible questions about the speed of the causal connection. Exactly when, relative to the distant killing, did Jocasta become pale? Faster than it would have taken light to bring the news of the event to her? If so, is the process Lorentz invariant or does it pick out some preferred reference frame?

So we skirt yet another quagmire of perplexities by simply omitting relational properties from our sufficient condition. The *local physical events* A and B are causally implicated with one another if B would not have occurred had A not (or vice versa). The notion of a local physical event need not be fully explicated, but locally observable changes, such as getting paler, setting off a photomultiplier tube, and rotating a filter, certainly count; becoming a widow certainly does not.

Finally, some remarks on the notion of causal implication are in order. Causal implication between events, as I intend it, is a very weak notion. We do not suppose that it follows from the fact that A is causally implicated with B that A caused B or B caused A. First, it is obvious that causal implication is a weaker notion than causation. Suppose, for example, that one is playing pool, trying to sink the 7 ball in the corner by a banked shot. Further suppose that in order to make the shot the ball must carom off a particular small section of the cushion. If the shot is made then the ball hitting that area of cushion is causally implicated with it falling into the pocket: if it hadn't bounced off that particular area it would not have gone into the pocket. But it is also arguable that the counterfactual dependence goes the other way around as well: if it hadn't gone into the pocket then it couldn't have hit the right section of the cushion. If it hadn't gone in it would have had to have missed the target area.[5] In any case, if it is at least sometimes correct to say that if the ball had not gone into the pocket it would not have hit the target area then causal implication can be symmetric between cause and effect while causation itself is not.

Not only is causal implication not strong enough to distinguish causes from effects, it also may obtain between events neither of which is a direct cause of the other. Immediately after the satisfying thunk of the 7 ball falling in the corner pocket, the cue ball drops dishearteningly into the side. In fact, any shot which manages to sink the 7 also scratches. So if the cue ball hadn't scratched, the 7 wouldn't have gone in either. Or so, at least, the expert might inform the amateur who laments "I would have won with that shot if only the cue ball hadn't gone into the side."

In this case neither of the two events, the cue ball and the 7 falling into their respective pockets, is a cause or an effect of the other. No mysterious causal process directly connects the happenings on different parts of the table. Still, given the disposition of the balls on the table, neither event would have happened without the other. The events are causally implicated with one another due to the operation of a common cause, viz., the interaction of the cue ball and 7 which sends each on its subsequent trajectory.

Local space-time events at space-like separation can routinely be causally implicated due to common causes and without the operation of any superluminal causal process. If we are to use the notion of causal implication to specify a sufficient condition for superluminal causation we must find a way to discount counterfactual connections secured by common causes. This is easily done.

Suppose all causal processes are constrained to propagate at or below the speed of light. Then any event would be causally influenced only by events in its past light cone and would only influence events in its future light cone. Furthermore, common causes for two events must lie in the overlap of their past light cones. In such a world, space-like separated events can only be causally implicated because of common causes. If A and B are space-like separated and A would not have occurred had B not occurred then there must also be some event in A's past light cone which would not have occurred had B not occurred, an event which may serve as the common cause of A and B.

We are now in a position to derive a *sufficient* condition for superluminal influences. If no causal influences are superluminal then it *cannot* be the case for space-like separated A and B that A would not have occurred had B not occurred *and everything in A's past light cone been the same*. By keeping everything in the past light cone the same we keep all causal influences the same in a causally local theory. The non-occurrence of B could not preclude A by having an influence on it. Nor could the non-occurrence of B be a reliable indication of the non-occurrence of A by being an effect of A since it does not lie in A's future light cone. Taking the contrapositive of the conditional given above yields our sufficient condition: given a pair of space-like separated events A and B, if A would not have occurred had B not occurred even though everything in A's past light cone was the same then there must be superluminal influences.

Satisfying this condition does not guarantee that B is a cause of A. As in all cases of causal implication, A might cause B, or B might cause A, or there might be a common cause. But since the common cause, if any, must lie outside of A's past light cone, all three possibilities involve superluminal influences.

The analysis just given is obviously modeled on Bell's account of local causal theories discussed in the last chapter. Bell asserted that in such a theory conditionalization on the intersection of the past light cones of two events should render the events statistically independent. We are imposing a more stringent criterion in our sufficient condition for superluminal causation. If Bell's condition fails then we have the following peculiarity: even after having taken account of *everything* in the past light cone of an event, the probability for the event is not uniquely fixed, for by taking into account some space-like separated event we alter our assessment of that probability. Bell's implicit argument is that the space-like separated event can only provide grounds for reassessing the probability if there is some direct or indirect causal connection, and since all causal connections mediated through the past light cone have been accounted for, the relevance of this distant event must depend on superluminal causal connections. Bell's argument is perfectly cogent and persuasive. But to satisfy the skeptical, our condition is even stronger: we require that even after taking account of everything in the past light cone, further conditionalization on the space-like separated event not only alters the probability assigned to A, it renders that probability either 1 or 0. Only in this case is it so that A would not have occurred if B had not.

Because of the intuitive connections between causation and counterfactuals we have labored to provide a sufficient condition for superluminal causation in terms of counterfactual conditionals. This may strike one as explaining the obscure through the more obscure: counterfactuals are notoriously slippery claims, vague, disputable, highly context-dependent. Indeed, it may seem that the evaluation of counterfactuals is not part of empirical science at all. Observation may inform us how the world is, but by what means could we determine how it would have been had it been otherwise? In the phrase of Asher Peres, "unperformed experiments have no results." Have we not left the precincts of physics for the unconstrained fantasies of speculative metaphysics?

A moment's reflection reveals such radical doubt about the scientific respectability of counterfactuals to be ungrounded. Determining what would occur under hypothetical circumstances is a central task of scientific practice. Cosmologists inform us that if the universe had a billionth of the mass it now has it would expand for ever, if it had a billion times its mass it would collapse. Empirical inquiry is used to determine if the Challenger would have exploded if it had been warmer on January 28, or if the inspectors had checked the O-rings. We turn to climatology to find out what the effects of a nuclear war would be. Beginning students calculate how frictionless objects would behave; graduate students how rotating charged black holes would distort the structure of space-time. The list could be extended ad nauseam (if it hasn't already been). The question, then, is not whether science deals in counterfactuals but rather how it manages to do so quite effortlessly, the supposedly problematic character of subjunctive conditionals notwithstanding. The answer to this question is quite simple and straightforward.

Scientific theories postulate laws which govern phenomena, laws like Newton's laws of motion and gravitation, or Einstein's field equations, or Schrödinger's equation. These laws state how physical magnitudes will, or could, evolve through time. These equations admit of different models which depict how the world might be. That is, the models depict the possible worlds allowed by the laws.

If we can specify a situation precisely enough, the laws will determine what would, or might, happen. Given the laws of gravitation, for example, and enough information about the masses and dispositions of the planets we can calculate how their orbits would be affected by the presence of another body of stated mass and location. Reasoning of this sort was used to discover the existence of Neptune and could equally well be used to determine how orbits would be perturbed by merely hypothetical planets, comets, etc.

Counterfactual claims can be implied by laws if the antecedent of the conditional is precisely enough specified. And if the laws are correct, if we have gotten the laws of nature right, then we can know about at least some unrealized possibilities. Given a set of laws we may be able to evaluate counterfactuals, and thereby to discern some causal connections. But now we appear to be stuck, for we have not yet supposed that we know what the laws of nature are. All we have supposed is that, whatever they are, they yield the predictions of quantum mechanics for correlated pairs of photons. But without *specific* laws we don't get specific counterfactuals, and without these we can't evaluate causal claims. So we must go beyond the correlations alone to consider the sorts of laws which could produce them.

Two World Pictures

If we do not have a particular set of laws before us, how can we proceed? Fortunately, possible laws fall into two general classes, and we have already discussed an example of each class. On the one hand, there are *deterministic* laws according to which neither of the measurement events in the photon pair experiment involves stochastic elements; the example of this sort of law is provided by Bohm's theory. On the other hand, there are *stochastic* laws according to which at least one of the measurement processes incorporates an irreducibly random element. Standard quantum mechanics, or quantum mechanics with modifications such as those suggested by Ghirardi, Rimini, and Weber (1986), fall in this camp. The analysis in each of these cases takes a rather different form, so we will begin by briefly reviewing the fundamental ontological picture provided by each.

According to Bohm,[6] the world consists at base of two sorts of objects. There is the quantum wave-function, which always evolves deterministically in accord with Schrödinger's equation. And there are particles which always have definite positions each of whose trajectory is a function of the wave-function and of the positions of the other particles. Given an initial wave-function, a Hamiltonian operator (which determines how the wave-function will evolve), and the positions of the particles, the laws fix a unique future evolution for the entire system.

According to orthodox quantum mechanics, in contrast, the wave-function provides a complete specification of the physical state of the world. There are no extra ("hidden") facts such as exact particle positions. This wave-function generally evolves deterministically in accord with Schrödinger's equation, but sometimes it changes in a different way, during wave collapse. The when and how of wave collapse form the nub of the measurement problem in quantum mechanics. Theories of all stripes have been forwarded: wave collapse occurs when a macroscopic change happens, or when a result is recorded, or when a conscious observer becomes aware of an outcome, or perhaps simply now and then at random. Fortunately the differences between these approaches need not concern us. We need only note that wave collapse, of whatever sort, introduces a stochastic element into the dynamics. A random choice is made, one which could have turned out either way given the laws of nature. Some sort of wave collapse (or non-Schrödinger evolution) is needed by any theory which holds the wave-function to be complete. Otherwise, as Schrödinger's cat shows, perfectly familiar experiments will not have definite results.

Some readers may feel that neither of our two exemplar theories – Bohm's theory and orthodox quantum mechanics with wave collapse – represent the right way to understand quantum theory. Physicists sometimes tend to be of two minds about the wave-function. On the one hand, the wave-function is held to be a complete description of physical reality: "hidden" variables are rejected. On the other hand, reduction of the wave-function is considered to be merely epistemic: it reflects a change only in our knowledge of the world, not in the world itself. Further, Schrödinger's equation is supposed to be universally valid, for few physicists are engaged in the project of modifying this fundamental dynamical law. The difficulty with this position is that it is self-contradictory. If Schrödinger's equation is universally valid and the wave-function is a complete description of physical reality then the cat simply does not end up either alive or dead. Wave collapse cannot be a mere artifact of coming to know whether the cat lived or died if the laws of physics imply that neither event occurred.[7]

For our purposes the question of how much the wave-function represents physical reality and how much is merely a function of our ignorance, and correlatively whether wave collapse reflects a physical or merely epistemic change, is irrelevant. Each reader may answer this question as he or she likes. But at the end of the day the question still remains: do the physical processes involved in measurement contain irreducible stochastic elements or not? According as the answer is given, one or the other of our analyses will apply.

Let's start with the deterministic case. Suppose the result of each observation is fixed by some set of physical facts together with the laws. Then we know that the physical facts which determine the outcome cannot lie exclusively in the past light cone of the measurement event. If they did, the outcome of each measurement would be determined independently of the setting of the distant apparatus since that setting (and all the events which determine the setting) may be at space-like separation. Each particle would then be disposed either to pass or be absorbed for each possible polarizer setting independently of the state of the distant polarizer. We know that no such situation can reliably recover the quantum correlations and violate Bell's inequality.

So in the deterministic case there must be at least some states in which the result of an observation on one side depends on the distant setting: that is, there must be at least some conditional instructions like those discussed in chapter 1. In terms of counterfactuals this means that in at least some cases had the distant polarizer been set differently the local result would have been different. But changing the distant setting either does not require changing anything in the past light cone of the local measurement or, if it does, this is already a case of superluminal causation.

Imagine that we set our polarizer based on radiation coming in from the Big Bang (figure 5.1). If the radiation had been different, the distant setting would not have been the same, and so (in at least some cases) the local result would have been different. The change in incoming radiation would either demand a change in the physical state of the past light cone of the local measurement or not. If not, then we have satisfied the condition for superluminal causation: the distant setting and local result are the required A and B. If on the other hand a change in the distant incoming radiation would already imply a difference in the past light cone of the local experiment then the first such difference is itself an example of superluminal causation.

It is instructive to see how this superluminal causation appears explicitly in Bohm's theory. Bell has worked out the details for the case of measuring spins of correlated electrons. The setting of each spin analyzer affects the Hamiltonian and hence the evolution of the wave-function. We have already

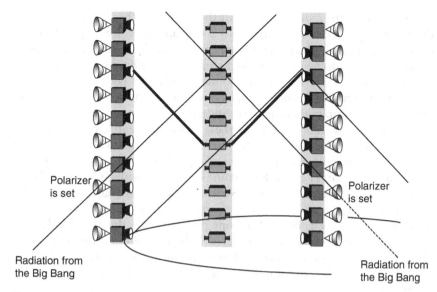

Figure 5.1 Ruling Out Subluminal Common Causes

seen in chapter 4 that this change does not effect the *statistical* predictions made by orthodox quantum mechanics. But in Bohm's theory, where the wave-function determines the trajectories of the particles, we can trace more than just statistics. In some cases the different wave-function which would have resulted had the distant spain analyzer been set otherwise would have sent the local particle on a different trajectory. Bell has derived the equations which demonstrate this dependency in Bohm's theory (Bell 1987, p. 132).

It has long been recognized that Bohm's model entails superluminal influences (indeed, Bell's work was in part inspired by the question whether such instantaneous connections were an inescapable consequence of any theory which could yield the predictions of quantum mechanics). This judgment itself confirms the adequacy of our analysis of causation. Bohm's theory, being deterministic, entails precise claims about how certain changes would have given rise to different outcomes, in particular about how changes in distant settings would lead to different local results. Any deterministic theory must have some such consequences.

The indeterministic case has seemed to be the main hope for avoiding superluminal causation. The general reason for optimism here is clear: in an indeterministic world the evaluation of counterfactual claims becomes problematic (see note 3). If a radioactive atom decays at 12:00 via a truly indeterministic process, what can be said about when it would have decayed had I been wearing a different colored tie? It might be misleading to say that it still would have decayed at 12:00 since that suggests that some antecedent factors destined the atom to decay just then. On the other hand, saying that the atom might not have decayed just then had I been wearing a different tie also can mislead: it carries the implication that the color of tie had some influence on the decay event. So in a stochastic world one must take great care either with the evaluation of counterfactuals or with the analysis of causation in terms of counterfactuals, else every event be adjudged a cause of every other simply in virtue of the indeterminism.

Fortunately, we can circumvent these difficulties entirely. We are now assuming that the laws are such that at least one of the two measurements involves an irreducibly random process, a process which (given everything in its past light cone) could have come out either with the photon being passed or with it being absorbed. Now consider the special case when the two distant polarizers are set in the *same* direction. Suppose both of the photons are absorbed. For at least one photon this is due to a fundamentally random process. What if that process had come out differently, if the photon had passed the polarizer? Then we know that its partner would have also passed since the laws imply a perfect correlation between the particles in this circumstance.

Since the laws imply perfect correlations between the photons when the polarizers are set in the same direction they also support the counterfactual claim that had the distant photon passed so would the local one have. The two events, the reaction of each photon to its analyzer, are causally implicated with one another. This alone, of course, does not require superluminal influences until one rules out the possibility of a common-cause explanation. Fortunately, eliminating common causes is even easier in the stochastic case than it was in the deterministic one. The fundamental feature of a stochastic process is that it could have come out differently even though *all* of its causal antecedents were unchanged. So any event which would have been different had the stochastic process had one of its alternative possible outcomes must be an effect, immediate or more distant, of that process.

Given stochastic laws, the counterfactual connection between the behavior of the two photons satisfies our sufficient condition for a superluminal causal connection. The indeterministic measurement process could have come out differently even though its entire past light cone were unchanged. And had it come out differently so would the other, space-like separated, measurement. This might be due to a direct causal connection between the two measurements or because the result of the distant stochastic process influences the physical state in the past light cone of the local observation. In either case superluminal influences are involved, for otherwise the only events which could counterfactually depend on the stochastic result would lie in the future light cone of that measurement event.

Without a more particular specification of the laws, we cannot discover exactly how the causal connection between the two events is secured, but just as Bohm's theory provides a useful illustration of the deterministic case, so the standard quantum-mechanical picture illustrates one way the stochastic scenario can be realized. If we assume that the wave-function provides a complete specification of the physical state of a system then the occurrence of particular results (such as passage or absorption) is inconsistent with the universal validity of Schrödinger's equation. The wave-function must occasionally change in a different way. This second sort of evolution is generally associated with measurements or observations, but it is the form of this second kind of evolution, not the circumstances which occasion it, which concerns us. (In the theory of Ghirardi, Rimini, and Weber (1986), for example, the second sort of evolution is not associated with measurement or observation *per se.*)

In the standard quantum picture the pair of photons are originally produced in a state in which neither has a definite polarization in any direction. In this state, each photon, when impinging on a polarizer oriented in any direction, would have an equal chance of passing the analyzer and of being absorbed. These probabilities reflect not merely epistemic failures on our part but rather a fundamental indeterminism in nature.

If a polarization measurement is made on one of the photons in the original state it will either pass the polarizer or not and, as a result, the state of the pair will be transformed. If, for example, a measurement is made on the right-hand photon with the polarizer set at 30° and the photon passes then the left-hand photon is no longer ascribed a state of indefinite polarization but rather a state of definite polarization at 30°. If the right-hand photon is absorbed then its twin acquires a definite polarization at at 120°, reflecting the fact that it will be absorbed if measured at 30°. So in the standard picture the physical state of the left-hand photon depends quite specifically on the result obtained on the right, and depends as well on what sort of measurement is made on the right. This dependency is strong enough to support definite counterfactual claims when the two polarizers are aligned, counterfactuals about how things would have gone differently on the left had they gone differently on the right.

It can become problematic to determine in detail how the physical state of one photon depends on the career of the other. In non-relativistic quantum mechanics the picture is clear: the particles begin in a state in which neither has a definite polarization, and stay in that state until the *first* photon reaches its detector. The observation of that photon initiates a stochastic process which eventuates either in the photon being absorbed or passing the polarizer. A collateral effect of this measurement is a change in the second photon from a state of indefinite polarization to a state of definite polarization in a direction either parallel or perpendicular to the direction in which the first is measured. This change in the second photon is simultaneous with the measurement on the first. The second photon, if its polarizer is set in the same direction as the first, then undergoes a *deterministic* evolution: its new state of polarization fixes whether or not it will pass. If measured at any other angle the second photon will undergo a stochastic interaction with the likelihood of its passage being determined by the angle between the orientations of the two polarizers.

In the relativistic domain this story is unintelligible. If the observations are made at space-like separation then there is no fact of the matter about which is temporally "first," and hence about which interaction is stochastic. Nor is there any unique notion of simultaneity which could fix when the distant partner's state changes, even if we could determine which photon undergoes the stochastic measurement.[8] So it is extremely unclear how the non-relativistic account of wave collapse is to be translated into the relativistic space-time. As of 2006, an explicit relativistic theory of this form exists. It is discussed in chapter 10.

The alert reader will doubtless have noticed that our argument that any *stochastic* theory will have superluminal influences has made no use at all of Bell's theorem. We have adverted only to the perfect correlation among photons when measured at the same angles; what happens when the polarizers are not aligned does not matter. Of course, perfect correlations alone do not imply superluminal causation. They do so *only* in the context of stochastic laws, for in this case one photon must be somehow sensitive to the outcome of the distant indeterministic process. The perfect correlations are easy to reproduce without superluminal influences in a *deterministic* theory: simply provide each photon with the same "instruction set" specifying how it is to respond to polarizers set in various directions. But the deterministic theories run afoul of other correlations, as we have seen.

Any attempt to avoid superluminal causation, then, is faced with a dilemma. If stochastic laws are proposed one is impaled on the perfect correlations. This danger can be avoided by deterministic laws, but then one is gored by Bell's inequality. Standard quantum mechanics embraces the first horn, Bohm's theory the second.

The architecture of this argument for superluminal causation explains some peculiarities surrounding the original reception of Bell's work. Bell's first proof examined whether any local deterministic hidden-variables theory could reproduce the predictions of quantum theory, answering in the negative. This hardly came as earth-shattering (or heart-breaking) news to most physicists since the search for deterministic hidden-variables theories was regarded as a pointless enterprise, motivated by nostalgia for classical physics and an inability to accept the fundamental chanciness of nature. If Bell showed that there must be something wrong with local deterministic theories, why not lay the blame at the doorstep of determinism rather than locality?

Bell, however, recognized his results as a refutation of *all* locally causal theories, and it is easy to see why. Stochastic theories involve an obvious element of superluminal causation. Deterministic theories were the only hope for locality from the beginning, a hope Bell extinguished. Bell was not studying deterministic theories out of a misplaced aversion to indeterminism, he was studying them as the only hope for retaining local (non-superluminal) causation.

But if the non-local character of stochastic theories such as standard quantum mechanics was so obvious simply from the perfect correlations, why wasn't this fact noticed long before Bell's work? And if noticed, why weren't people *worried* about the non-locality? Why wasn't there a concerted effort to find a local (and hence deterministic) replacement for quantum mechanics? The answer to the first question is that the non-locality of quantum mechanics *was* noticed, and pointed out, by Einstein, Podolsky, and Rosen. The answer to the second question is much less clear, involving as it does one of the darker episodes in the history of modern physics.

The EPR Argument

If the quantum mechanical description of our pair of photons is complete, i.e. if it represents (directly or indirectly) the entire physical state of the photons, then polarization measurements must involve some fundamentally stochastic process, for the state does not ascribe a definite polarization to either photon, and the theory allows that photons in such indefinite states may either pass or be absorbed by a given polarizer. But when the two distant polarizers are aligned the passage or absorption of each photon depends on the passage or absorption of its twin. We have just argued that this dependency constitutes a causal connection. Hence if the quantum mechanical description of the photons is complete, the measurement carried out on one photon influences the state of the other.

Put the other way around, if there are no superluminal influences, if the physical state of each photon is unchanged by observations carried out on its distant partner, then the quantum mechanical description of the system cannot be complete. This second formulation of the point was argued by Einstein, Podolsky, and Rosen (EPR) in their paper (1935). EPR did not use our example of photon polarization (or the other modern favorite: electron spin) but rather constructed a quantum mechanical state in which the *positions* of two particles would be precisely correlated. The relevant state does not ascribe a definite position to either particle, but does predict an exact *distance* between them if both positions are measured. Hence by measuring the position of one particle the position of the other can be exactly predicted. After one measurement is made, the wave-function collapses to a new state which assigns a *definite* position to the remaining particle.

We can now raise the familiar dilemma. Does the wave collapse represent a real change in the physical state of the second particle or not? If it doesn't, then since after the collapse the particle has a definite position, so too it must already have had one independently of the distant measurement being performed. But then the original quantum state was incomplete since it did not represent either particle as having a definite position. If, on the other hand, wave collapse does represent a real physical change caused by the distant measurement then there are superluminal influences.

The EPR argument could end at this point. Einstein and his collaborators thought the second option, wave collapse reflecting a physical rather than epistemic change, to be obviously unacceptable. Hence they concluded that the quantum mechanical description is incomplete and that the correlation between the observed positions of separated particles could only be explained if each particle always had a definite position revealed by the measurement. Only thirty years later did Bell show that this approach also cannot avoid superluminal causation. The local deterministic theories which recover the EPR correlation must violate Bell's inequality.

Unfortunately the crisp and powerful central argument of the EPR paper was obscured by an unnecessary bit of grandstanding (probably due to Podolsky – see Fine 1986, ch. 3). For not only does the quantum mechanical state constructed in the paper allow one to predict the position of the distant particle by measuring the local one, it also allows one to predict the *momentum* of the distant particle by measuring the *momentum* of the local one (the net momentum of the pair is zero). Hence, by parity of reasoning, the distant particle must also always have a determinate momentum, supposing superluminal effects to be disallowed. So the distant particle must have a determinate position *and* momentum, but *no* quantum mechanical state can represent a particle with determinate position and momentum. So the physical state of the particle eludes representation in purely quantum mechanical terms. (Translating the example to photons, quantum spinors cannot represent a photon having determinate polarization in more than one direction at a time. Let "polarization at 0°" and "polarization at 45°" stand in for "position" and "momentum" respectively to formulate the EPR argument in our terms.)

This extra twist of the knife sinks the previously simple EPR argument into the muddy waters of modality. The first argument focuses attention directly on the use of wave collapse, squarely facing the status of the wave-function and its dynamics. The second argument adds the complications surrounding counterfactuals, for in no case can one be in a position to predict with certainty both the position and momentum of a particle. But consider the case where we actually measure the local particle's position and so ascribe a definite position to its distant partner. The argument runs: we *could have* decided to measure the local particle's momentum, thereby fixing the twin's momentum. But if the local measurement does not influence or change the distant particle's state then the momentum of the distant particle cannot depend on which measurement we decide to carry out. The distant particle always must have a definite position and momentum, *neither* of which is reflected in the initial wave-function.

On the surface, this new argument, which appeals to possible but actually unperformed measurement, appears to commit a modal fallacy. It seems to mistake the uncontroversial claim that if a momentum measurement had been carried out there would have been a result (i.e. some result or other) for the controversial assertion that there is a *particular* result which would have occurred if a measurement had been made. The latter is true only under the assumption of determinism: the physical state of the particle together with the laws of nature must guarantee that interaction with a measuring device could have but one unique outcome. This tempts the philosopher to the satisfying pronouncement that although Einstein *et al.* were splendid physicists, they unfortunately had not studied enough modal logic. (In recent debates the so-called "principle of counterfactual definiteness" has been the center of a similar debate about the assumptions needed to derive Bell's inequality. See Redhead 1987, pp. 90ff.; Stapp 1989.)

An argument can be made that even the more complex EPR argument does not commit any modal fallacy. Under the assumption that measuring one particle does not influence the other, *only* a deterministic theory can account for the correlation, for in an indeterministic milieu there would be no way for the distant particle to "track" the result of the local position or momentum observation. And given a local deterministic theory no modal fallacy is committed: there is a unique outcome which would have obtained had a position (or momentum) measurement been made on each particle. But this whole issue is of secondary importance: the heart of the EPR argument lies in its searching examination of the status of wave collapse in orthodox quantum mechanics, not in the attack on Heisenberg's uncertainty relations.

How was the EPR argument received? The "official" reply was penned by Bohr shortly after the publication of the original EPR paper (Bohr 1935). Unfortunately, Bohr's response escapes any clear interpretation. In order to fully appreciate the provenance, and ultimately the inadequacy, of Bohr's reply a brief digression on his notion of complementarity is in order.

When wrestling with the notorious wave/particle duality in quantum mechanical descriptions, as well as with the associated uncertainty relation between incompatible observables, Bohr hit on the idea that certain families of concepts can be coherently applied to a physical system only when the attendant environmental or experimental circumstances permit. For example, in order to speak meaningfully of the position of a particle there must be some rigidly fixed physical reference frame with which the particle interacts.

To speak of its position in the absence of such a frame is akin to asking what time it is somewhere on the Sun: although the question may appear well formed (as well formed as asking what time it is in Kiev), the framework required for it to have content simply does not exist. Bohr then went on to note that certain of these frameworks may be mutually incompatible. For example, whereas the notion of position requires a rigid physical frame (with reference to which the position is determined), the concept of momentum demands instead that the particle interact with a freely moving structure which can recoil. Such pairs of concepts which demand incompatible conditions for their valid application were denominated "complementary." In particular, the conditions needed for a coherent space-time description of an object were deemed incompatible with the conditions required by a description in terms of energy and momentum. (Note that this is not really a wave/particle duality: classical particles have *both* position and momentum.)

Whatever their final adequacy, Bohr's ideas on complementarity may seem to have internal coherence and plausibility when applied to a *single* particle. They are not strictly consistent with the usual interpretation of the quantum state, according to which a system has a determinate value for a quantity whenever it is an eigenstate of the associated observable, irrespective of the environment in which it finds itself.[9] Still, the fundamental idea that the prerequisites for the applicability of a "space-time" description of events may be incompatible with those presupposed by a "causal" (i.e. energy/momentum) description is passably clear.

Against this background the pronouncements of Bohr on the EPR argument assume a recognizable shape. Bohr begins by seemingly agreeing with EPR that a *causal influence* of the local measurement on the distant particle is inadmissible: "of course, there is in a case like that just considered no question of a mechanical disturbance of the system under investigation during the last critical stage of the measuring process" (1935, p. 699). Nonetheless, Bohr insists that there *is* a sort of influence of the local measurement on the distant state of affairs:

> Even at this stage there is essentially the question of *an influence on the very conditions which define the possible types of predictions regarding the future behavior of the system.* Since these conditions constitute an inherent element of the description of any phenomenon to which the term "physical reality" can be attached, we see that the argumentation of the mentioned authors does not justify their conclusion that quantum-mechanical description is essentially incomplete. (1935, p. 699)

Bohr's idea is evidently that the distant experimental arrangement, the observation of the distant particle, plays not a causal but a conceptual role in the description of the local situation. Somehow, just as the presence of a rigidly fixed physical body plays an ineliminable part in providing the notion of position with an unambiguous meaning, so the experimental procedure performed on one particle determines, or restricts, the very concepts which can be meaningfully applied to the other.

Although we can make sense of the general drift of Bohr's remarks, careful consideration shows them to be completely inappropriate to the EPR experiment. Perhaps we can make this clearer by translating the problem to the case of photon polarization.

The claim that the photon on the right is polarized at 0° will certainly be well defined if a polarizer on the right is set to 0°: passage or absorption by this polarizer would play the same role for polarization as interaction with a rigid frame plays for position. (Again, it is notable that orthodox quantum mechanics does not require the presence of an actual polarizer to "make sense" of claims about polarization: the photon is so polarized if it is in the eigenstate of the appropriate operator.) No reference, implicit or explicit, need be made to *distant* experimental arrangements or to the outcomes of *distant* experiments. Further, we may imagine that *both* sides of the experimental apparatus be fixed well in advance of the experiment being performed: both polarizers are set at, say, 0°. Now the "conditions which define the possible types of predictions regarding the future behavior of the system" are perfectly determinate. We are interested simply in whether the photon on the right will pass its polarizer or not.

The original quantum state assigned to the photons, the state analogous to that which EPR set out to prove incomplete, ascribes no definite polarization to either photon, and hence makes no definite predictions about passage or absorption. But once the left photon actually passes, that state is changed to one in which the right-hand photon *has* a definite polarization and will definitely pass. So the question is not what role *experimental arrangements* on the left have in defining quantities on the right but rather how one is to understand the *change in state* on the right which follows the *outcome* on the left. If no physical change has occurred, then the photon on the right which has a definite, well-defined polarization after the left-hand measurement, must also have had one before. So the original quantum state was incomplete.[10]

In sum, EPR had a perfectly cogent argument against the completeness of quantum mechanics if one grants that distant measurements cannot affect the physical state of a local system. Further, under the same assumption they could show that the underlying true dynamics must be deterministic in order to account for the observed perfect correlations. The need for determinism arose not from some metaphysical bias but from the predicted (and now observed) facts. Anyone hoping to avoid superluminal influences would have to construct a deterministic "hidden-variables" theory. Bell's work, finally, demonstrated the untenability of local deterministic theories, making the acceptance of "spooky action at a distance" unavoidable. Still, for at least thirty years the physics community preferred to rest content with the unintelligible pronouncements of Bohr rather than facing the consequences of Einstein's critique.

A Note on Wave Collapse

According to the foregoing analysis, the fundamental point of the EPR paper lay not in the attempt to discredit the Heisenberg uncertainty relations (interpreted ontologically) but in the careful consideration of the implications of wave collapse. In orthodox quantum mechanics, wave collapse is at least superficially a source of potential conflict with Relativity. In non-relativistic quantum mechanics it is treated as an instantaneous process: in virtue of a measurement performed in one place the wave-function everywhere undergoes a sudden radical change.

Wave collapse is not treated as a physical process in most expositions of quantum mechanics. Indeed, it is hardly treated at all. One is most tempted to regard it as a purely epistemic rather than physical change: we collapse the wave-function by conditionalizing on newly discovered facts just as Watson adjusted his subjective probabilities for the claim that Holmes' body lay at the foot of Reichenbach falls when Holmes whisked off his bookseller disguise. At that moment no peculiar physical change took place at the falls, all that changed was Watson's representation of what might be there.

We have already noted that the idea (1) that wave collapse is merely epistemic is inconsistent with other positions commonly held, viz., that (2) the wave-function completely specifies the physical state of a system, that (3) the wave-function always evolves in accord with Schrödinger's equation, and that (4) Schrödinger's cat ends up either completely dead or completely alive. But more can be said than just this. (2), (3), and (4) alone are inconsistent since Schrödinger evolution leads to a wave-function which no more represents the cat as alive than it represents it as dead. (1), (2), and (3) are also inconsistent since, if we assign the correct initial state to a system, Schrödinger's equation alone would generate the complete physical state of the system at any later time. Hence once we calculate the wave-function there would be no further "facts" which we could come to know, upon which we could conditionalize. This highlights the fact that (2) and (3) alone are inconsistent with the idea that there are any stochastic processes in nature, for Schrödinger evolution is perfectly deterministic.

In short, if we want to hold that wave collapse is merely epistemic, or that nature is indeterministic, or that the cat either ends up dead or ends up alive, then we must abandon either (2) or (3). Attempts have been made in both directions. (The most well-known attempt to retain both (2) and (3) is the many-worlds interpretation discussed in note 1 of the Introduction.)

If one is to hang on to the completeness of the wave-function then some means must be introduced whereby the wave-function can evolve only into states in which detectors either fire or fail to fire and in which cats are either determinately alive or determinately dead. This means taking some non-Schrödinger evolution seriously as a *physical* process. Proposals of this sort have mostly been rather vague, being suggestions about the circumstances in which non-Schrödinger evolution occurs. The most detailed proposals have been put forward by Ghirardi, Rimini, and Weber (1986) and Philip Perle (1986).

The key requirement for all of these proposals is that the non-Schrödinger evolution cannot (usually) take place before either photon reaches its detector: only the peculiar "entangled" quantum state predicts perfect correlations, so the disentanglement provided by wave collapse cannot happen while both particles are in flight. So all of these schemes require exactly the superluminal causal connection we have noted. Somehow interaction with one of the measuring devices leads not to a superposition of the photon being absorbed and the photon passing but rather to one or the other result. And the state of the distant photon changes correlative to this result so as to guarantee correlation if its polarizer happens to be at the same orientation. In short, if the wave-function is complete then wave collapse reflects a physical change by which the state of a photon is changed by events at space-like separation.[11]

The other possibility is to retain Schrödinger evolution but insist that the wave-function does not represent all of the physical facts in the world. This route is followed by Bohm. He allows the cat's wave-function to evolve into a state which is a superposition of live cat and dead cat. But the cat is not made up of wave-function: it is made up of particles with definite positions. Those positions will either be those of the parts of an active, noisy cat or of an inert, silent, former cat.

The key to understanding "wave collapse" in Bohm's theory lies in the observation that, for the most part,[12] if the particles are in the "live cat" configuration then their future trajectory will depend *only* on the "live cat" piece of the wave-function, and if they are in the "dead cat" configuration only the "dead cat" part of the wave-function matters. Hence one can, for all practical purposes, throw away the irrelevant part of the wave-function once one has determined how the particles actually stand. Throwing away the irrelevant part is wave collapse, a purely epistemic act of practical approximation. The (as it were) uninhabited part of the wave-function is always there, but for macroscopic systems it is unlikely to have any observable effects. This is not so for small systems or for specially prepared states of large systems. Most famously, in the two-slit experiment even though the particle goes through only one slit its trajectory is significantly affected both by the part of the wave-function associated with that slit and by the part of the wave-function associated with the other slit.

So if one wants wave collapse, the non-Schrödinger "evolution" of the wave-function, to be merely epistemic then there must be more to the world than is represented in the wave-function. But ironically Bohm's theory, which does away with instantaneous wave collapse as a superluminal *physical* process, has been most severely criticized for postulating superluminal influences. This is the price of clarity. By being explicit about what exists and how things interact Bohm presents a definite object of analysis. One can calculate how changes on one side of the apparatus lead to effects on the other. Most other "interpretations" of quantum mechanics glory in vagueness. The exact status of the wave-function and the nature of wave collapse are not made quite clear. The wave-function is at once both physically real (to explain interference effects) and merely epistemic (to avoid wave collapse as a real physical process and other embarrassments, see note 7). When questions of ontology become acute the wave-function can be robbed of all physical significance, becoming a mere "calculational device."

For our purposes none of this matters. Let the wave-function be a mere calculational device, let wave collapse be just a formal step in a mathematical procedure devoid of any direct physical significance. All that we have had to do with the calculational device is calculate with it. The device tells us that when the polarizers are aligned, had the photon behaved differently on the one side so would the other, distant photon. If we add to this that one of the photons behaved as it did through sheer chance, by a truly stochastic process, then we know that it could have reacted differently even though all of its causal antecedents had been the same. And so we get a counterfactual-supporting causal connection between the photons which cannot be explained by a common cause.

It is fortunate that quantum mechanics predicts perfect correlations between photons when the polarizers are aligned. This allows us to state in detail how gross observable events would have been different on one side had they been different on the other. It may be felt that this heavy reliance on perfect correlations constitutes a flaw in our argument, for no *actual* experimental apparatus will display such perfection. Polarizers, detectors, sources all can be made only with a finite degree of precision. Suppose, for example, that in an actual experiment the polarizers are misaligned by $1°$. Then even when both photons pass we cannot say with certainty what would have happened if one had been absorbed. The best we can say is that the other would have had a 99.97 percent chance of absorption. Our definite counterfactual connection fails.

Two sorts of answers to this criticism are appropriate. First, it seems perfectly fair to discuss what a theory tells us about what would happen in various idealized situations. Perhaps as a practical matter we can't build perfect polarizers or align them exactly. Still, the theory informs us what would happen if we could and all of our empirical experience supports these claims. Improved experimental techniques yield closer and closer approximations to the ideal limit.

Second, any theory which takes stochastic laws seriously at the ontological level must take ascriptions of probability equally seriously. If we believe that a photon approaching a polarizer has a 50 percent chance of passing and a 50 percent chance of being absorbed, and that these probabilities are reflections not of our ignorance but of a basic indeterminism in nature, then we must take an event's having a particular probability as a basic *physical* fact. In this case a change from 50 percent probability of passage to a 99 percent probability is a *physical* change. And if we know that had photon A been absorbed (rather than having passed) photon B would have had a 99 percent chance of absorption (rather than a 1 percent chance) then we know that A's passage is causally implicated with B's physical state.

An event's having a certain probability, or having its probability change, is not directly observable. Hence one may be leery of taking changes in probability as real physical changes. But here one should at least take a clear stand. Like Bohm, we can declare that all of the probabilities of quantum mechanics are merely epistemic: the laws of nature are actually deterministic. Or one can embrace fundamentally probabilistic laws and accept physical propensities are real and irreducible. But to do both and neither, to float between the epistemic and non-epistemic interpretations without a principle or a guide, choosing each option as seems most convenient at the moment, is to shirk one's intellectual responsibilities. And either way one settles on in the end, superluminal causation follows as a consequence.

But is it Causation?

Our strategy so far has been to provide a sufficient condition for causal connection in terms of counterfactuals and to evaluate the counterfactuals by means of laws or law-schemas. I have argued that if one of the measurement processes is indeterministic then there must be a causal connection between the outcome of that measurement and the outcome of the distant measurement. But some authors have suggested that although some connection exists between the space-like separated events, that connection ought not to be called causal.

The issue is often put in terms of a supposed contrast between two sorts of superluminal connection, sometimes unhelpfully called "action-at-a-distance" and "passion-at-a-distance." (The "at-a-distance" part does not connote discontinuity but rather "at space-like separation.") Or again, it is claimed that although quantum theory postulates some sort of superluminal connection, nonetheless a "peaceful co-existence" between quantum theory and Relativity prevails. The general idea is that although action-at-a-distance would entail a very severe clash with Relativity, passion-at-a-distance is relativistically acceptable (Redhead 1989, p. 150) or anyway less unacceptable (Shimony 1990, p. 40, but see also p. 309).

The distinction between action and passion is drawn differently by different authors, as we will see. But before plunging into textual exegesis we should pause to note the peculiarity of the whole exercise. After all, if there are conflicts with Relativity one would suppose that they arise from the "at-a-distance" part, whether the superluminal connection be active or passive. If one takes the speed of light to be a limit then it ought to limit all physical processes, not just "actions." If the heart of relativity is not any limiting role of the speed of light, if it is instead, say, Lorentz invariance, then it must be shown why superluminal actions are any more of a difficulty than superluminal passions.

In some cases the distinction between action and passion collapses into a contrast that we have already discussed. In Abner Shimony's work, for example, the contrast between action and passion is assimilated to that between violations of parameter independence vs. violations of outcome independence, the latter being claimed to allow the peaceful co-existence with Relativity (1989, pp. 29ff.). This judgment of compatibility arises from an undefended interpretation of Relativity as forbidding only superluminal signals and from an acceptance of Jarrett's flawed argument that any violation of parameter independence would allow such signals. (Recall that Bohm's theory violates parameter independence but does not allow any practical means of sending signals since the precise value of the hidden variables cannot – in principle – be controlled or ascertained.) We have already seen that Relativity does not *per se* forbid superluminal signaling. It here remains for us only to determine whether suitability for use in a practical signaling scheme should be taken as a necessary condition for postulating a causal dependence.

In the orthodox interpretation of quantum mechanics the perfect correlation between distant photons (when the polarizers are aligned) cannot be used to transmit information since neither of the two connected events can be controlled. It is evident that *if* one could induce a photon to either pass the polarizer or to be absorbed then superluminal signals could be sent.[13] For if perfect correlations continue to hold then modulating the behavior of one of the photons would produce a corresponding pattern in the observable behavior or its twin. So the question arises: how essential is controllability for the very notion of causality? If controllability is built into causation then the no-signaling theorems immediately convert into no-causation proofs.

To begin, *practical* controllability clearly plays no part in the concept of causation. Earthquakes, for example, are paradigm causes of manifold effects but lie entirely outside of human control. No method for using earthquakes to send signals has ever been proposed but we would not downgrade their causal status on that account.

It may be urged that this is a mere technical deficiency. After all, it is at least *physically* possible to control an earthquake, and with sufficient ingenuity and expenditure of capital no doubt a system for relating the results of, say, papal enclaves by means of temblors could be devised.

Harder cases can be imagined. After a certain point the collapse of a star into a black hole can be prevented by no physical means. Or, since it costs us nothing to be grandiose, consider the Big Bang. It is surely *physically* impossible that the Big Bang could have been controlled or used to send messages but it nonetheless has *effects*.

Physical considerations, even considerations of principle, may still leave the philosophical mind unimpressed. It is at least *metaphysically* possible to control the Big Bang and use it to send messages. God might decide, for example, to employ the following scheme: if He is forgiving the universe will be open, ever-expanding; if He is jealous the universe will be closed (and if He is malicious we won't be able to tell whether the universe is open or closed). But now the point of the whole exercise becomes obscure. In whatever sense the Big Bang is metaphysically controllable, so too is the passage of a photon through a polarizer. The suggestion that for a connection to be causal a divinity must be able to use it to send messages is hardly a useful analysis.

The temptation to use these fantastic theological scenarios does not arise from any deep conceptual connection between causation and signals. The scenarios rather illustrate a *counterfactual* connection: had the Big Bang been different so would the later course of the universe have been. Whether the Big Bang could – in any physically or metaphysically interesting sense – have been controlled is neither here nor there. In general if one adds control of one variable to a counterfactual-supporting connection one gets signaling,[14] but the addition is strictly irrelevant to the existence of the causal connection.

Michael Redhead has argued that yet another feature of causation is missing from the quantum connection, a feature which (it seems) has nothing to do with signaling. Redhead denominates this condition *robustness*. Since the idea is rather slippery I can do no better, to begin, than to quote him at some length:

> By robustness of a causal relation we mean the following: A stochastic causal connection between two physical magnitudes a and b pertaining to two separated systems A and B is said to be *robust* if and only if there exists a class of sufficiently small disturbances acting on $B(A)$ such that $b(a)$ screens off $a(b)$ from these disturbances.

Denoting the disturbance acting on B by d, then the first part of this condition can be rendered formally as

$$\exists D \left(\forall d \in D \left(\text{Prob} \left(a = \varepsilon_a / b = \varepsilon_b \& d \right) = \text{Prob} \left(a = \varepsilon_a / b = \varepsilon_b \right) \right) \right)$$

A similar condition can be written down for a disturbance acting on A.

The requirement of robustness as a necessary condition for causal relation means that suitably small disturbances of either relata do not affect the causal relation. This is essentially the basis for the mark method for identifying causal processes. The processes propagate small disturbances (marks) in a local event structure in accordance with the causal law at issue. (1987, p. 102–3)

Redhead's exposition contains both a formal condition and an informal gloss. Let's start with the formal condition, identifying the exact form it would take for our photons.

I have claimed that there is, in the orthodox interpretation of quantum mechanics, a causal connection between the passage of one photon and the passage of its twin. So let A stand for the system (photon plus polarizer) on the right and B stand for the system on the left. a and b are variables representing the interaction of photon and polarizer and each can take on either of two values, namely passage or absorption. The term ε_a stands for the specification of the outcome in system A and similarly for ε_b. Instances of the formula Prob $(a = \varepsilon_a/b = \varepsilon_b)$ are, for example, Prob(right photon passes/left photon passes), the probability that the right photon will pass given that the left one does. When the polarizers are aligned these conditional probabilities are all either 0 or 1, when misaligned the probabilities take on intermediate values.

The formal condition written down by Redhead therefore amounts to this: there should be some class of statements D such that for any statement d in that class the likelihood that the right photon passes given that the left photon passes *and* d is equal to the probability that the right photon passes given only that the left photon passes.

The formal condition cited above is therefore trivial to satisfy. Let D, for example, be a set of statements including the assertions such as that Quayle was Vice-President or that Saturn has rings or, if you will, that the price of tea in China has gone up. Conditionalizing on this further information will change the probabilities not one whit. Indeed, the only way that the condition could be violated is if any bit of information, no matter how far-flung or miscellaneous, altered our assessment of the likelihood of the right photon passing given that the left one has. It is hard to know what we would say in such a case. The world would be supercausal, every event connected intimately to every other, and perhaps the very notion of causation would collapse. But this condition is not violated by our photon, or by anything else in the universe.

In order for the formal condition to be of any use at all there must be some restriction on the set D. Redhead calls the d's *disturbances*, and it is unlikely that anyone would consider the price of tea in China a disturbing factor in our experiment. But how are we to get guidance on what is to count as a disturbance?

In a footnote to another article on this topic, Redhead cites David Lewis and Brian Skyrms for notions "related" to his:

> Lewis maintains that in a causal relation between events it is necessary that the events could have happened in a different way, in particular at different times, while the truth of the causal statement is upheld. Skyrms has the idea of resilience of a conditional probability in a stochastically causal situation, which has its value unaltered by conditionalization on further information, in particular on how the events were brought about. In our discussion we consider just the case where events consist in magnitudes taking values. (1989, p. 148)

How does the formal condition on causation fare if disturbances are interpreted in this way?

No better. It is, for example, irrelevant *when* the left-hand photon passes its polarizer. The conditional probability for the right-hand photon passing is unchanged whether its partner passed at 12:00 or 12:05 or a week ago last Tuesday. Similarly, it doesn't matter how the passage was brought about. It is irrelevant, for example, what sort of polarizer was used, what its temperature was, etc. According to Lewis' or Skyrms' "related" notions we still have a causal connection.

So of what sort of disturbances does Redhead actually make use? Unfortunately, the discussion here lapses into a rather technical mode. Disturbances are represented by the "coupling of particle B to uniform *c*-number fields of arbitrary strength" (1987, p. 103). It is then shown that after having been coupled to such a field the conditional probability considered above is altered.

Passing over for a moment the implausibility of the idea that "coupling to a uniform *c*-number field" could be part of any necessary condition for causation, what does Redhead's observation amount to? The physical situation is this (making adjustments for the fact that Redhead discusses electron spin rather than photon polarization). Certain substances (called "optically active") possess the ability to rotate the direction of polarization of plane-polarized light. If a beam of light polarized at, say, 0° is passed through such a substance it will emerge polarized at a different angle, for example at 5°. Now suppose we let one of our correlated photons pass through an optically active substance (or alternatively let a spin-correlated electron pass through a magnetic field). The correlations between the photons will thereby be changed. For example, whereas originally the photons would be certain to yield perfectly correlated results only when the polarizers are aligned, now they would be perfectly correlated only when the polarizers are misaligned by 5°. The quantum state of the pair of photons has been changed due to the rotation of the state of polarization of one particle.

Not surprisingly, the presence of an optically active substance alters our conditional probabilities. With the substance in place the probability that the right-hand photon will pass at 0° given that the left one has drops from 100 percent to 99.24 percent. It is just as if instead of the photons being rotated the polarizer were rotated in the opposite sense by 5°. So if we restrict the d's to be only statements about the presence of optically active substances then the formal condition fails. But it is hardly surprising that such interference with the photons changes the correlations and hardly convincing to claim that this fact debars the two sides from exhibiting a causal connection.

Redhead's formal condition for robustness looks like a non-starter and the plausible patches for it offer no improvement.[15] So let's leave it behind and consider a slightly different line of argument.

In Redhead's analysis the non-robustness of the link between the two photons arises simply and purely from the fact that the link depends on Ψ, the quantum state of the particles. Any change in Ψ results in a change in the correlations between the two sides[16] and Redhead admits as a disturbance *only* interactions which change the relevant part of Ψ. But once we know Ψ no further information will change our assessment of the conditional probabilities. So, asks, Redhead,

> if nonrobustness arises simply from the dependence of $\mathrm{Prob}^{|\Psi>}(a = \varepsilon_a/b = \varepsilon_b)$ on $|\Psi>$, why not restore robustness by regarding $|\Psi>$ as contributing common cause of the correlation? But there's the rub. If $|\Psi>$ is a partial common cause, then it must be overlaid by a direct causal link, since for fixed $|\Psi>$, $\mathrm{Prob}^{|\Psi>}$ $(a =_a/b =_b)$ has a robust form as a function of the values ε_a and ε_b. But a direct causal link is unacceptable on relativistic grounds. (1989, p. 149)

In short, if robustness is the mark of causation it is there for the taking: all one needs to do to have a robust statistical link is take account of the quantum state of the particles! And after all, how could one expect there to be *any* calculable correlation between the events if the quantum state is *not* taken into account? There is no general correlation between one photon passing a polarizer and some other, randomly selected photon doing so: only pairs in the entangled state exhibit the correlation with which we are concerned.

But Redhead has tried to stack the deck against robustness since, on his analysis, it is a necessary condition for causation. He throws out the quantum state as a possible factor simply because including it would yield a causal link between the sides. But even this rather desperate move won't do the trick. For we have a *robust* correlation between the two photons if we only conditionalize on the nature of the photon source and the presence of any optically active substances (or sources of *c*-number fields), omitting any *explicit* mention of the wave-function. Providing just this information about the conditions of the experiment allows us to calculate conditional probabilities which will be unaffected by any further information. This is, after all, just what we do: we assign a Ψ-function on the basis of the experimental design, and use it then to derive the probabilities.[17] Redhead tries to cast doubt upon using the Ψ-function as an appropriate object for conditionalization by questioning its ontological status, but the status of the macroscopic equipment used to realize the experiment (and to produce any disturbing conditions) is unassailable.

Having dismissed these objections to calling the quantum connection causal we at last obtain a positive result. Reliable violations of Bell's inequality need not allow superluminal signaling but they do require superluminal causation. One way or another the occurrence of some events depends on the occurrence of space-like separated events, and the dependence cannot be understood as a result of the operation of non-superluminal common causes. When we discussed superluminal signals and energy transport we asked purely hypothetically about their compatibility with the relativistic space-time structure. Now we have a non-hypothetical question: does the fact that the world contains superluminal causal connections contradict the Theory of Relativity?

Superluminal Influences and Relativity

Much of our discussion about the compatibility of superluminal causal connections with Relativity can be taken over directly from our examination of superluminal signals. Signaling between two places requires at least the possibility of a causal connection between them since the state of the receiver must nomically depend on the state of the transmitter. So if superluminal signals are compatible with Relativity so *a fortiori* are superluminal influences.

As with superluminal signals, it is often asserted without further justification that Relativity forbids superluminal influences (e.g. Butterfield 1989, p. 114). Sometimes the supposed prohibition on action-at-a-distance is simply taken to be equivalent to the prohibition of superluminal signals, eliding the gap between causation and signaling (e.g. Jarrett 1984). So let's briefly review some possible objections to an event having effects at space-like separation.

First the analytical or conceptual objection: cause and effect always occur in a determinate time order, cause preceding effect. But space-like separated events don't have a determinate time order since different reference frames disagree about which came first. Therefore superluminal causation is impossible.

We have already seen the major premise of this argument to be mistaken. Circumstances in which an effect precedes its cause can be coherently described when there is some intrinsic mark of one of the relata (such as controllability) which identifies it as the cause. In the case of the quantum connection (in the orthodox interpretation) such controllability is absent, so neither event can in this way be assigned the role of cause. There are two possible responses to this situation. One is to recognize causal *connections* between events which don't support any distinction between cause and effect. The events are counterfactually connected in the right way for causation but we simply forgo any further specification of the relation.

The other possibility is to insist that in such circumstances temporal order comes to *define* which is cause and which is effect, the earlier event being the cause. Here is one way such a proposal might be implemented. In non-relativistic quantum mechanics wave collapse is taken to be instantaneous, occurring at the moment of measurement. The most obvious relativistic simulacrum of instantaneousness is simultaneity-in-a-reference-frame. So suppose every observer adopts the following rule: the entangled quantum state (in which neither photon has a definite polarization) changes to a state in which both photons have definite polarization at the moment (in my reference frame) when the first measurement is made ("first" in my reference frame). Then different observers will tell different stories about any given run in the Aspect experiment.

If the polarizers in figure 5.2 are aligned and both of the photons pass their polarizers then observer A (using unprimed coordinates) will tell the following tale. The photons were produced in the entangled state. When the right-hand photon reached its detector an irreducibly stochastic choice was made. The photon, which could just as well have been absorbed, passed its polarizer. At that moment (t = constant) the state of the left-hand photon abruptly changed from having no definite polarization to being polarized in the direction of the right-hand polarizer. When it reached its polarizer a bit later it was certain to pass, already being in a eigenstate of polarization in that direction. No stochastic choice is made at the left-hand measurement.

Observer B, using the primed coordinates, gives a symmetrical but opposite account with "right-hand" and "left-hand" exchanged. In this reference frame the left-hand photon arrives first, undergoes the stochastic process and initiates the wave collapse.

Both stories agree that there is a causal connection between the two sides, but they disagree about which event should count as the cause and which the effect. If both stories can ultimately be considered equally valid we will have to accept that unequivocal causal connections do not require unequivocal identification of the cause and the effect. The tenability of this method for representing wave collapse will concern us in the coming chapters.

The second objection we must consider rests on the supposed impossibility of causal loops. The chain of reasoning goes: (1) superluminal causation implies (in at least some frames of reference) backwards causation, that is, later events influencing earlier ones; (2) if backward causation were possible one could arrange for physical situations with no acceptable outcomes. One could, for example, set up a gun to fire at a target and connect the target by backward causal links to the gun in such a way that the gun fires if and

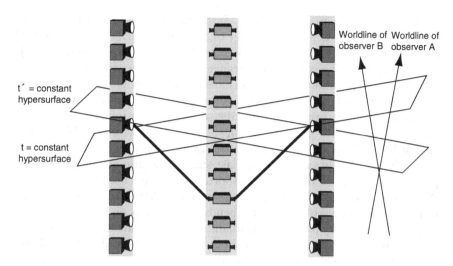

t′ = constant
hypersurface

t = constant
hypersurface

Worldline of Worldline of
observer B   observer A

Figure 5.2   Aspect's Experiment and Two Observers

only if the target is not hit. It appears that the gun can neither fire nor fail to fire; so (3) something in the description of the situation must be wrong. Superluminal causation is the prime suspect.

This argument fails at every step. First, superluminal causation does not imply that an identifiable cause precedes an identifiable effect in *any* frame of reference, as we have seen. For appropriately symmetrical connections each frame could use its own judgment of time order to decide which event is cause and which effect.

Second, superluminal causation does not *per se* imply the possibility of causal loops. The last chapter already provided examples of superluminal signals which do not permit signal loops. But the obstacles to having causal loops are even higher. In order to get signals into one's own past light cone at least *two* space-like signals must be chained together, as in figures 4.10 and 4.11. For signals, such chaining is always possible: a signal from P to Q can be retransmitted from Q to R to give a signal from P to R, for by definition the transmitter of a signal is controllable and the reception of the signal observable. One need only construct a coupling at Q which will pass along the signal received from P.

But the quantum mechanical causal connection is *not* of the controllable variety. So the existence of causal links from P to Q and from Q to R does not imply the possibility of a causal connection between P and R. If we cannot *control* at Q the Q → R connection or cannot *observe* at Q the P → Q effect then no effective linkage between the two causal processes can be made.

To be concrete, suppose in the set-up of figure 4.11 one correlated pair of photons will be measured (with polarizers aligned) at P and Q while a second pair is measured at Q and R. Further, suppose it is somehow established that in the P–Q pair the passage or absorption at P causes the passage or absorption at Q and in the Q–R pair the passage or absorption at Q causes that at R. We still cannot arrange for a causal correlation between P and R since we cannot make the behavior of the Q–R photons contingent on the behavior of the P–Q photons. An observer at Q can see how the first pair behaves but is unable to force the second pair to follow suit. Causal paradoxes are usually generated by *signal* loops and where causal processes cannot be used to send signals paradoxes cannot arise.

Finally, we should note that the last step of the argument from causal paradox fails even if all the earlier shortcomings have been patched. John Wheeler and Richard Feynman have pointed out that in any realistic situation where the nature of the effect depends continuously on the nature of the cause the supposedly paradoxical situation has a consistent resolution. This solution may be odd and unexpected but it will be in accord with all of the physical laws (Wheeler and Feynman 1945).[18]

Superluminal influences in relativistic space-time entail neither insuperable conceptual problems nor causal paradox. But we still may have a fundamental conflict with Relativity if the influences require that some reference frame be intrinsically preferred over all others. That is, we still have to face the problem of Lorentz invariance: can the causal law be formulated in a Lorentz invariant way?

The parallel discussion of superluminal signals in the last chapter cannot simply be repeated without modification for superluminal causation. We found that if ubiquitous superluminal signals are allowed then either signal loops are possible or else some unique global reference frame must play a privileged role in the theory, breaking Lorentz invariance. But since uncontrollable causal connections cannot be woven into causal loops this dilemma no longer applies.

As encouraging as these observations are, they do not discharge all of our worries. In orthodox quantum theory the causal connection between photons is secured, in part, by wave collapse. We will examine the possibility of a relativistic theory of wave collapse in chapter 10. In Bohm's theory the connection derives instead from the way the wave-function guides particle trajectories. No relativistic version of Bohm's theory exists. So the project of reconciling the superluminal influences of quantum theory with Lorentz invariance demands serious attention, and we will turn directly to this task in chapter 7.

We have so far been able to go a great distance while remaining very vague about the details of the physical laws which govern the pair of photons. It is fortunate that a causal connection between the two sides of the apparatus can be demonstrated both on the assumption of stochastic and on the assumption of deterministic laws. But we have now gone as far as we can using only the requirement that the observable correlations predicted by quantum mechanics be recovered. We must begin to speculate about exactly how the superluminal connection might be made.

Notes

1   Some of this work is done in my "A Modest Proposal Concerning Laws, Counterfactuals, and Explanations" (2007).

2   If the run is infinitely long, the probability of perfect match tends to zero. But zero probability is not impossibility: perfect match could occur even in an infinite run, and any particular sequence of matches is no less likely than any other specific result for an equally long run. Besides, we never have infinite runs of any sort, so it is unclear why the case should concern us.

3   The status of counterfactual claims in this context is a matter of some dispute. Suppose we flip two independent, irreducibly stochastic coins and both come up heads. What should we say about what would have happened if the first had come up tails? On the one hand, we may argue thus: since *ex hypothesi* there is no causal influence of one coin on the other, a different result for one would have made no difference to the other; it would still have come up heads. David Mermin calls this the *Strong Baseball Principle* (1989, p. 42). On the other hand, one can argue that since the second coin is governed by an irreducibly stochastic process we can't say what would have happened in other circumstances: it might have come up heads and might have come up tails. So we cannot say that had the first come up tails the second would still have come up heads. This sort of analysis is defended by Michael Redhead (1987, pp. 92ff.). A framework in which both of these intuitions can be captured is sketched in my "A Modest Proposal" (2007). The important point for us here is that on *neither* view is it correct to say that had the first come up tails so would have the second.

4   In John Bell's terminology, Jocasta's widowhood is not a local beable.

5   It sounds a bit odd to say that if it hadn't gone in it would have missed the target area, but the reasons for this oddness are not immediately obvious, so I confine my conjectures to this note.

   David Lewis (1979) argues that a "backtracking" counterfactual such as "If the ball had not gone in it would not have hit the target area" would be considered *false* under the "standard" resolution of vagueness of counterfactuals, and he proposes a general theory which is designed to yield this result. Lewis does come up with a theory according to which, in most cases, had an event at time t been different still all of the events up until t would have been the same. Unfortunately, Lewis' theory also implies that under the standard resolution, "If the ball had missed the pocket, a miracle would have occurred" is true! And although it may sound odd to say that the ball hitting the target counterfactually depends on it going into the pocket, it is rather worse to say that the universal validity of the laws of nature counterfactually depend on the ball going into the pocket. Lewis, in an attempt to avoid the implication that present events can influence the past, falls into the absurdity that every present event (in a deterministic world) influences the laws of nature (and, incidentally, causes the absence of past miracles).

   In the end, it seems much more reasonable to trace the oddness of "If the ball had not gone into the pocket it would not have hit the target" not to its falsity but to the prosaic fact that counterfactuals concerning datable events *usually* are framed in terms of the dependence of the later on the earlier. Thus the usual locution at least conversationally implies that the antecedent took place before the consequent. One uses the clearer "If A had occurred, B would have had to have happened" to avoid this conversational implication, not to invoke a non-standard way of resolving vagueness. This is especially important since the cause/effect relation is usually time asymmetric, so the usual locution places the cause in the antecedent and the effect in the consequent. If this account is correct, Lewis' attempt to derive the direction of causation from general considerations about the semantics of counterfactuals is hopeless.

6   This is Bohmian particle mechanics. Bohmian field theory is similar but with the field values playing the role of particle positions.

7   An example of the peculiar doublethink prevalent among physicists is provided, in an oddly inconclusive passage, by David Mermin:

> We know what goes through the two slits: the wave-function goes through, and then subsequently an electron condenses out of its nebulosity onto the screen behind. Nothing to it. Is the wave-function itself something real? Of course, because it's sensitive to the presence of both slits. Well then, what about radioactive decay: does every unstable nucleus have a bit of wave-function slowly oozing out of it? No, of course not, we're not supposed to take the wave-function that literally; it just encapsulates what we know about the alpha particle. So then it's what we know about the electron that goes through the two slits? Enough of this idle talk: back to serious things! (1990a p. 187)

8   It is sometimes said that in the original entangled state neither photon *has* a physical state since the state of the pair cannot be expressed as the product of states of the two photons. If so, one might be reluctant to call the change which results in the second photon having a definite polarization a change in its physical state: antecedently it (considered on its own) didn't *have* a physical state. Nonetheless, going from *not having a physical state to having a physical state* is some sort of change, call it what you will!

9   It might also be noted that the doctrine of complementarity makes only tenuous contact with the quantitative details of the uncertainty relations, details which derive from features of the wave-function itself.

10 Nor can one remedy the situation by claiming that quantum mechanics is not *per se* incomplete but rather that the quantum state originally assigned to the photons was wrong. First of all, quantum mechanics asserts the possibility of such a state and implies that the state will give rise to perfect correlations. Second, *only* the "entangled" state, in which neither photon has a definite polarization in any direction, predicts perfect correlation in all directions.

11 In the theory of Ghirardi, Rimini, and Weber the collapse takes place independently of any "observation" or "measurement" or mentality. Hence in their theory there is *also* an epistemic change which occurs when we *find out* that a collapse has happened. But the original physical collapse is itself still a superluminal process.

12 The qualification must be added since *in principle* it would be possible to detect interference effects due to the existence of both parts of the wave-function, although for any normal macroscopic object the obstacles to arranging for such interference are practically insurmountable.

13 See Jones and Clifton (1993) for an extended discussion of this point.

   It may be worth noting that superluminal signals could also be sent if we could enhance our power of observation. If we could determine exactly what state a single photon is in, then we could use orthodox wave collapse to send messages. For if I set my polarizer at 0° then after wave collapse the distant photon will either be in an eigenstate of polarization at 0° or of polarization at 90°. If I set my polarizer at 45° then the distant photon will end up in an eigenstate of either 45° or 135°. So if I could determine the state of the photon I could send messages in the orthodox theory *by manipulating my polarizer*. An unsuccessful scheme of this sort was proposed by Nick Herbert (see 1985, ch. 8). This illustrates how little it means that orthodox quantum theory does not violate so-called parameter independence: even in that theory how I set my parameter influences the physical state of the distant photon.

14 This is not universally valid, as both Bohm's theory and orthodox quantum theory show. In Bohm's theory, control of one causally important variable (the polarizer setting) does not give practical signaling due to lack of control over the initial state. In orthodox quantum theory control over such a variable (again polarizer setting) does not give signaling since the relevant effect (the polarization state of the distant photon) cannot be observed.

15 Nancy Cartwright and Martin Jones (1991) provide a more exhaustive search through possible interpretations of Redhead, and they also fail to find a convincing argument. Richard Healey (1992) considers the interpretation of quantum connections on several analyses of causation, including Redhead's. Healey thinks that one cannot unproblematically judge the connection to be causal, chiefly because no spatio-temporally continuous process connects the measurements. I do not find it at all plausible that spatio-temporal continuity is a necessary condition for causal connection: classical action-at-a-distance was thought to be unmediated but preeminently causal.

16 Strictly speaking this isn't true. Redhead's Ψ is *only* the spin state of the electrons; he omits the part of Ψ that describes the *positions* of the particles. Disturbances in this part of the wave-function (for example by accelerating or decelerating the particles) would not alter the spin correlations.

17 Oddly, Redhead seems to deny this truism.

Robustness says that if *a* and *b* are two physical magnitudes which are causally related then the cause *b* screens off *a* from sufficiently weak disturbances *d* acting on *b*. But if this is violated, why not incorporate *d* as part of the cause of *a*? But this resets the problem. We now require that *b* and *d* jointly screen off *a* from whatever brought about *d*. Call this *d′*. Continuing in this way, I claim that if incorporating *d*, *d′*, *d″*, … etc. as part of the cause never succeeds in restoring robustness, then we have no justification in talking about a lawlike connection at all. (1989, p. 151)

But even acceding to Redhead's most restrictive terms, incorporating d (the c-number field) *does* render the form robust: how the field is *produced* is quite irrelevant. If quantum mechanics did present a situation like that sketched by Redhead it is hard to see how one could make any predictions at all.

18 See also Clarke (1977). An extended discussion of this issue can be found in Maudlin (1990), which would be highly recommended except that its conclusion is incorrect. (For those interested, the bullet can hit the edge of the target directly above 0°, partially activating the magnets, which would then carry the gun only part way on its trajectory toward 180°.) Wheeler and Feynman's account holds for all continuously operating devices which can be built in a simply connected space-time.

6

Secret Messages

A pair of photons created together in the entangled state and allowed to separate remain, no matter how great the separation, causally connected to one another. If the quantum correlations are to be reproduced the reactions of one photon to polarization measurement must be sensitive to the type of observation carried out on its distant twin. The exact nature of this causal connection cannot be inferred solely from the observable correlations.

In chapter 1 we set the problem of reproducing the quantum correlations as a practical puzzle, a game to be played by a pair of people attempting to replicate the statistics displayed by the photons. We saw that the task cannot be accomplished without information about a distant measurement being available to at least one particle. In order to perform the trick the two photons, or the two players, must remain in communication. We have now reached a point at which we can be more precise about the nature of the "communication." The communicative channel between the distant particles does not employ the transmission of energy. It cannot be used by outsiders to send superluminal messages of their own. But still the incidents which befall one of the pair have an influence on the other.

One measure of the communication which flows between two points is provided by information theory, a measure well suited to our purposes. Information transmission is not defined by reference to energy transfer, or signaling ability, or even directly by reference to actual causal processes. Rather, the information communicated from a transmitter to a receiver is a measure of how much one can infer about the state of the transmitter (and with what confidence) given the state of the receiver. We may use a familiar

Quantum Non-Locality and Relativity: Metaphysical Intimations of Modern Physics, Third Edition. Tim Maudlin.
© 2011 Tim Maudlin. Published 2011 by Blackwell Publishing Ltd.

anecdote to illustrate how information can flow between two points when nothing physical is exchanged between them.

In "Silver Blaze," Holmes makes the following remarkable observation while discussing the case with Watson:

> "Is there any point to which you would wish to draw my attention?"
> "To the curious incident of the dog in the night-time."
> "The dog did nothing in the night-time."
> "That was the curious incident," remarked Sherlock Holmes.

Holmes, it later appears, has been able to deduce from the silence of the dog that the horse, Silver Blaze, must have been abducted by a familiar member of the household; otherwise the dog would have given the alarm. The dog's not barking provides Holmes with information about the goings-on in the stable on the fateful night.

Although information is transmitted from the stable to Holmes via the dog's quiescence, evidently nothing else is. No energy or matter need pass from the crime to the detective nor any causal process link them. So information can be very ghostly indeed. The information is transmitted only because of the *modal* structure of the situation: had a stranger been present the dog would have been heard. (Of course, in this case the information is transmitted only because it was at least *possible* for energy to have emanated from the dog and arrived at the ear. Similarly, although information may not require any *actual* causal process between two events, there must be at least a *possible* nexus of influences connecting them if the state of one is to provide information about the other.)

Trading in information allows us to operate at a high level of abstraction, ignoring the exact details of the communication channel. We can sensibly ask how much information must flow between the wings of our experiment if the quantum statistics are to be generated, leaving aside questions about how the communication is brought about. In the sequel we will take advantage of this studied agnosticism concerning mechanism and simply pursue the question of how much, at a minimum, one photon needs to know about the measurement carried out on the other. In particular, we will consider the *setting* of one polarizer as a source and the *state* of the distant photon as the receiver, and ask how much information must pass from one to the other in order that the quantum predictions be recovered.

Limits for Uncommunicative Partners

Let's return to the game of chapter 1 to pursue this question further. In that game you and your partner were allowed to collaborate on a strategy for answering questions. After separating, each of you would be asked one of

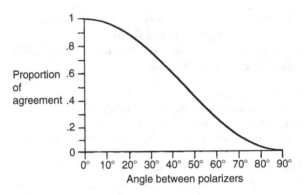

Figure 6.1   The Quantum Correlations

two questions: "0?" or "30?" on one side, "30?" or "60?" on the other. You must answer either "Passed" or "Absorbed." The task is to devise a strategy which returns the quantum statistics: 100 percent agreement when both are asked "30?"; 75 percent agreement when the questions differ by 30; 25 percent agreement when the difference is 60. We saw that in the absence of any information about the question asked of one's partner no solution is possible.

We will now expand the game to include the whole set of possibilities which confront the photons. After the photon pair is created each member could in principle be passed through an analyzer oriented at *any* angle. The "question" asked of you and your partner could be any number between 0 and 180. And after a long series of runs one must find that when the difference in angle between questions asked in α, the proportion of agreement is $\cos^2 \alpha$. Figure 6.1 illustrates the target pattern. Suppose, to begin, that no communication is allowed between the partners. How close to the target pattern can one come?

Let's start with two facts. First, the answers must agree whenever the same question is asked on both sides, no matter what the question may be. Second, when the questions differ by 90° the answers must differ.

As we noted in chapter 1, employing stochastic mechanisms for answering questions only makes the situation *worse* for the two players since more information (the result of the stochastic process) must be communicated between the two sides. Without this information the players cannot possibly be sure to agree when the same question is asked. Since we are trying to minimize the amount of communication needed, we will confine ourselves to deterministic strategies.

Partners unable to communicate can satisfy the first constraint, perfect correlation when the polarizers are aligned, only by adopting identical strategies for answering questions. The partners must therefore be provided with

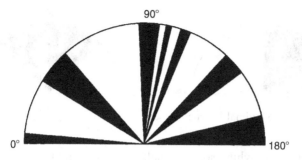

Figure 6.2 An Instruction Set

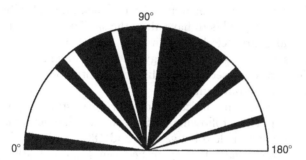

Figure 6.3 Satisfying the First Two Criteria

identical instructions for answering all possible questions. We can represent the instruction set as a semicircle with some segments colored black. If the polarizer is set at an angle whose corresponding point is colored black then the photon will be absorbed, otherwise it passes (figure 6.2). Of course, the instruction set may be varied from run to run.

To satisfy the second criterion, perfect disagreement if the polarizers are misaligned by 90°, every black segment on the semicircle must have a corresponding white segment displaced by 90°. Hence we need only specify the coloring of the region between 0° and 90°: the area between 90° and 180° will just be its inverse image (figure 6.3). It follows that half of the semicircle will be colored white, half black. Only instruction sets constructed in this way can satisfy the first two requirements.

Having taken care of the cases of perfect alignment and perfect misalignment of the polarizers we must now consider all of the intermediate cases. There are several ways of proceeding, and we will follow one of the most obvious. Let's consider the behavior of the photons when the polarizers are only slightly misaligned, when α is near zero. As figure 6.1. shows, there remains a very high degree of correlation in these cases, so nearby regions on the instruction set should generally agree in color. The region surrounding a point colored black should itself tend to be colored black, the region

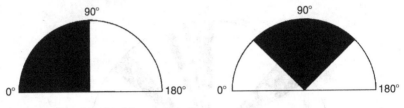

Figure 6.4 Two Maximal Instruction Sets

near a white point white, so that small differences in angle will not usually lead to differing responses.

Given that half of the semicircle must be black and half white the best one can do in this regard is to have all of the similarly shaded regions contiguous, a solid quarter-circle of white and a solid quarter-circle of black. Two examples of this maximal strategy are shown in figure 6.4. (The white region in the instruction set on the right is one continuous piece since 0° and 180° are identical "misalignments.")

In the absence of communication the very best strategy you and your partner can adopt is the following. When you are still together in the room, randomly choose an angle ϕ between 0° and 180° (with all angles equally likely to be chosen). Paint the region between ϕ and $\phi + 90°$ (mod 180°) on your instruction sheets black, that is, agree to answer "absorbed" if the question asked falls within that region and "passed" otherwise. Using this strategy each of you will answer "passed" approximately half of the time in the long run. There will be no correlation between the angle of either polarizer and passage or absorption at that polarizer. When the same question is asked on both sides you will always agree. When the questions differ by 90° you will always disagree. And when the questions are close to each other, e.g. when "50?" is asked on one side and "51?" on the other, you will give the same answer as often as you possibly could (in the absence of communication).

But just how often will you agree in this last case? The probability that you will *disagree* is just the chance that a black/white border in the instruction set lies between 50° and 51°. More generally, the probability that the questions "ψ?" and "$\psi + \alpha$?" get different answers is just the chance that such a border lies between ψ and $\psi + \alpha$. But this last probability is just proportional to α. If the slice between ψ and $\psi + \alpha$ is twice as large then the likelihood of catching the border in the region is twice as great. (This holds as long as $\alpha \leq 90°$, after which point one must worry about catching two borders.)

So the probability of disagreement scales linearly with the angle of misalignment from 0 percent disagreement with no misalignment to 100 percent disagreement for 90° misalignment. This response function is shown in

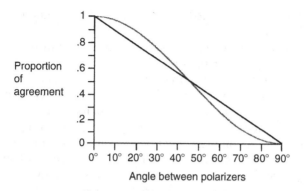

Figure 6.5 Best Strategy with No Communication

figure 6.5, along with the target curve. The linear relation shown is the very best anyone can do if there is no communication between the two sides of the experiment. As has been often noted, one hallmark of the entangled systems is that as the polarizers become misaligned the particles hold on to the correlations more strongly than any pair of causally isolated particles could. They show more correlation at small angles than is possible with no communication given 100 percent agreement for no misalignment and 100 percent disagreement when the polarizers are at right angles.

How Much Does a Particle Need to Know?

In order to improve upon the linear approximation of figure 6.5, either you or your partner must know something about the other's apparatus. The vital bit of information is the angle at which the distant polarizer is set. Just how much information about the setting on the distant wing is needed to be able to match the target curve?

To answer this question we first need some means of measuring information. The standard measure is the *bit*: one bit of information corresponds to a single binary digit, or to the answer to a yes/no question. A signal which transmits one bit of information allows the receiver to infer which of two possible states the transmitter is in. Two bits of information allow discrimination between four states, three bits among eight states, and so on. So let's start by asking how much your situation would improve if your partner could send one bit of information after you have been separated.

In terms of our game, imagine that after having left the room and before you have to answer the question posed to you, you receive a message from your partner containing one bit of information, a slip of paper containing either a 1 or a 0. Which message is sent may depend on the question which has been asked of your partner. How could you best make use of this new resource?

The way to take advantage of this new source of information is to have two instruction sets rather than one. To maximize the difference between using one of the sets rather than the other, the two sets ought to be as different as possible (with respect to the placement of their black/white borders). The most extreme difference obtains between pairs whose black and white regions are displaced by 45°, such as the pair depicted in figure 6.4. Let's call the instruction set on the left *Plan 0* and the instruction set on the right *Plan 1*. Your partner has the option of using either Plan 0 or Plan 1 and, by informing you of the choice, assuring that you use the same scheme. This provides a means of overcoming the problem of small angles.

The linear correlation of figure 6.5 fails to match the target in part because it requires more disagreement at small angles of misalignment than the photons display. This happens because when the polarizer settings Ψ and $\Psi + \alpha$ are close to each other the probability that a black/white border will fall between Ψ and $\Psi + \alpha$ is too high. Disagreement is likely to occur when the boundary in the instruction set happens to fall near to the direction of both polarizers. But using the one bit of information your partner is now able to avoid this danger. If her polarizer is set very near a black/white boundary for Plan 0 she can opt instead for Plan 1, thereby moving the boundary approximately 45° away from the polarizer setting. Suppose, for example, that your partner's polarizer is set at 91°. If you both follow Plan 0 your partner will answer "passed." If your own polarizer happens to be set at 89° you will answer "absorbed," giving a disagreement at small angle of misalignment (2°). By choosing Plan 1, though, this hazard disappears. Your partner answers "absorbed" at 91°, and Plan 1 assures that for any small angle of misalignment, indeed any misalignment up to 44°, you will agree.

Consider, then, the following strategy. You and your partner, while still in the room, randomly choose a direction ϕ between 0° and 180°. Plan 0 puts the black/white border at the chosen direction, Plan 1 offsets the boundary from this direction by 45°. When your partner is asked to respond she chooses the plan whose borders are *farthest* from her polarizer setting and communicates that choice to you using one bit of information. You adopt the corresponding plan when answering your question. What will be the result?

Since the boundaries in Plan 0 and Plan 1 are offset by 45°, the chosen plan will have borders at least 22.5° from the direction of your partner's polarizer. So if you and your partner always use this strategy you will *always* agree when the two polarizers are misaligned by less than 22.5°. Beyond 22.5° the rate of agreement drops linearly to perfect disagreement at 67.5°. If the angle of misalignment is between 67.5° and 90° you are certain to

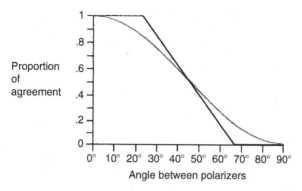

Figure 6.6 A One-bit Strategy

disagree with your partner. The overall pattern of correlation is shown in figure 6.6, along with the target curve.

Using only one bit of information we have overcome, even over-compensated for, the problem of small angles. We have devised a strategy which shows even higher degrees of correlation at small angles than do the photons, and more disagreement when the polarizers are misaligned by nearly 90°.

It may be useful to discover the precise content of that one bit of information. Ideally, each player would like to know exactly which question has been asked of his partner, exactly how the distant analyzer is set. But since that question can be any real number between 0 and 180 (the polarizer can be set at any angle), knowing the precise question would require the transmission of an infinite amount of information. That is, if one had that information one would be able to determine which of an infinite number of distinct possible situations obtains at the other end of the experiment. Using the one-bit scheme much less information about the state of the distant polarizer is sent. For example, suppose Plans 0 and 1 of figure 6.4 are used. Then your partner will adopt Plan 0 if her polarizer is set anywhere in region marked 0 in figure 6.7, and she will adopt Plan 1 if the polarizer is in region 1. The single bit of information has reduced our ignorance about the distant setting from total uncertainty to uncertainty about its location within the indicated region.

Our one-bit strategy does not match the target curve exactly, so we still have some work to do. Perhaps some combination of the one-bit and zero-bit (linear) strategies can give an improved approximation. But instead of working with so impoverished a palette, we should first note that the one-bit strategies described above is but one of an infinite family of one-bit strategies available to us, each of which yields a distinct correlation pattern.

When designing our first one-bit strategy we maximized the difference between the placement of the boundaries in Plan 0 and Plan 1. If we adopt

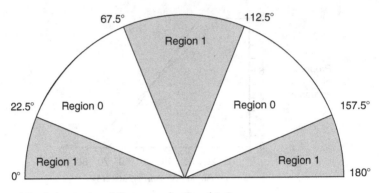

Figure 6.7   Informational Content of a One-bit Strategy

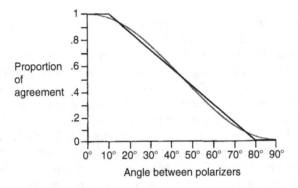

Figure 6.8   Another One-bit Strategy

less extreme measures we get less extreme results. Suppose, for example, that the black/white boundary of Plan 1 is shifted only 20° from the orientation in Plan 0 rather than 45°. In that case the plan whose boundary is farthest from the polarizer angle might be only 10° away. Perfect correlation would only be guaranteed for small angles from 0° to 10°, with the probability of agreement falling linearly from perfect correlation at 10° to perfect disagreement at 80°. (Alternatively, one could still use the original plans of figure 6.4 but only switch to Plan 1 if the distant polarizer is set within 10° of the boundary in Plan 0.) If one used this new one-bit strategy over a long series of runs the resulting correlation pattern would be that depicted in figure 6.8.

In general, any one-bit strategy with an offset angle of 2θ between Plan 0 and Plan 1 will give rise to a graph which is flat from 0° to θ and drops off linearly until reaching 0 at 90 deg $-\theta$. The zero-bit strategy is the limit as θ approaches 0: the simple linear relation. In the sequel we will refer to any strategy which produces a graph flat from 0° to θ and linearly dropping off

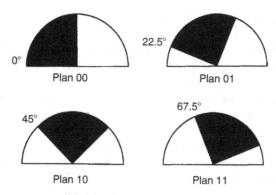

Figure 6.9   Plans Used in a Two-bit Strategy

to 0 at 90 deg $-\theta$ as a θ-*strategy*. Having the ability to communicate one bit of information makes available to us all of the strategies from the linear 0°-strategy to the maximal 22.5°-strategy.

How nearly can we match the target curve by using a mix of one-bit strategies? Surprisingly well: we can exactly match the target curve in the region from 0° to 22.5° (and, by symmetry, from 67.5° to 90°). This perfect match results from using the probability density $\rho(\theta) = (\pi - 4\theta)(1 - 2 \sin^2 \theta) = (\pi - 4\theta) \cos 2\theta$ for choosing among the strategies between 0 and 22.5 (θ in the probability density has been expressed in radians rather than degrees).[1] The probability density fixes the likelihood of choosing the various strategies: the probability of choosing a strategy between θ and $\theta + \delta$ is $\rho(\theta)\delta$ (for δ very small). The probability of choosing a strategy with $0 \le \theta < 22.5$ deg is

$$\int_0^{\pi/8} (\pi - 4\theta) \cos 2\theta \, \alpha\theta \cong 0.848$$

The remaining 15.2 percent of the time the strategy 22.5° is chosen.

The result of this mix of strategies comes very close to the target. But after an exact match between 0° and 22.5° there is a linear regime from 22.5° to 67.5°. Linearity cannot be avoided in this region if one uses only one-bit strategies since *all* of those strategies are linear between 22.5° and 67.5°. So our mix of one-bit strategies, although an excellent approximation, fails to give a perfect match to the target.

In order to extend the region of perfect match further more information must be sent. If your partner can send two bits of information then she can choose between four possible plans of action, such as those illustrated in figure 6.9. One of the plans can always be chosen so that the black/white border lies no nearer than 33.75° to the orientation of your partner's polarizer, thereby extending the flat part of the response curve to 33.75°. Using the same probability density we can now obtain a perfect match with the target curve from 0° to 33.75° using one-bit strategies 84.8 percent of the time, two-bit strategies (between 22.5° and 33.75°) 13.2 percent of the time, and the 33.75° strategy the remaining 2 percent of the time. The result of this mix is a curve which is linear only between 33.75° and 56.25°. Two-bit strategies send enough information to localize the distant polarizer setting to within a region which occupies a quarter of the original semicircle.[2]

Further improvements can be made in the same way. Absolutely perfect match to the target requires using three-bit strategies 1.74 percent of the time, four-bit strategies 0.22 percent, five-bit strategies 0.028 percent and so on, without end. To *exactly* match the target, your partner must in principle be capable of sending messages of unbounded informational content, but will be required to send complex messages only rarely. In the long term, to achieve the behavior that photons display requires that your partner transmit an *average* of 1.174 bits of information per experiment. A slightly more interesting number (as we will see) is the average value of 2^N, where N is the number of bits sent in a run. This represents the average for the number of states of the distant polarizer that one can distinguish between on the basis of the message sent, and comes out to 2.408. You and your partner must begin by choosing a particular strategy, with the probability for each strategy determined by the probability density $\rho(\theta)$. Most of the time you will choose a strategy which can be implemented using one-bit messages, but occasionally much longer messages will be needed.

Evaluation of Results

The question of how much information transmission is required to reproduce the predictions of quantum theory is an area of active research, and the results reported above are now obsolete. One extremely efficient method, capable of replicating spin measurements carried out on electrons using only one bit of transmitted information, has been discovered by B. F. Toner and D. Bacon (2003). The mathematical problem for this case is slightly different than for our polarizers because the spin can be measured in any direction in space. Bacon and Toner's method is important because the method never requires more than one bit: exactly one is sent on every trial. It is possible that even this quantity of information could be reduced.

As Bacon and Toner remark, the only proven bound on the amount of information needed to reproduce the predictions of quantum theory (and violate Bell's inequality) is due to Bell: the amount must be greater than zero. Given this key result, the significance of finding the scheme that requires minimal information is a bit hazy. If there must be a physical mechanism that transmits information from one side to the other, it is not clear that a method using less information would be more physically plausible than one using more. But the search for the most efficient scheme is fascinating, and it is possible that discovering the lower bound might open up some insight into the physics.

The conclusion of our investigation is rather equivocal. Given that the vital unknown, viz., the setting of the distant polarizer, can only be precisely conveyed using messages of infinite information content, it is remarkable that the predicted behavior violating Bell's inequality can be replicated using only a single bit. But the critical distinction in nature is not between a little information conveyed and a lot: it is between some and none. If there is a non-local information channel at all, it is unclear that it would cost nature more trouble to send many bits rather than fewer. Indeed, it is unclear that it is easier to find a theory in which only one bit of information is sent than a theory in which an infinite amount is sent.

In fact, this is precisely what happens in the standard (non-relativistic) quantum mechanical picture of wave collapse. Each photon starts out in a perfectly indefinite state of polarization. When the first photon is measured the wave-function changes so that the second photon becomes definitely polarized *either parallel or perpendicular to the angle at which the first was measured*. So wave collapse transmits an infinite amount of information. If we could see the state of the second photon after the collapse, determine exactly what its state of polarization is, we could infer the orientation of the distant polarizer (or, more accurately, we could infer that either the distant polarizer is set at a particular angle or perpendicular to it). Of course, the exact polarization state of the incoming photon is not observable. In this case it is the "observable effect" clause of the definition of a signal which prevents wave collapse from being a signaling process, since the setting of the distant polarizer does constitute a controllable state.

In the internal economy of photons standard wave collapse transmits a mixture of information about how the distant polarizer is set and about the result of the stochastic part of the distant measurement process. If the right-hand photon after the collapse is polarized at some angle θ then we know that either the distant polarizer was set at θ and the distant photon passed or that the distant polarizer was set at $\theta + 90°$ and the distant photon was absorbed. If we already know how the distant polarizer is set then the state of the local photon could inform us whether its partner passed its polarizer or not. We used exactly this connection in the last chapter to demonstrate a superluminal causal connection in stochastic theories. Some of this hidden information can be made available to experimenters: if we align the polarizers as in the EPR set-up then the passage or absorption of one photon provides accessible information about the behavior of its twin.

If our task is to exactly reproduce the predictions of quantum mechanics, the average value of 1.174 bits of information sent per experiment, or Toner and Bacon's constant value of 1 bit, have no clear practical significance: the superluminal information transmission, in whatever quantity, must exist. In standard quantum theory, wave collapse transmits unbounded amounts of information. But if we relax the conditions of our game in another way this number, or similar averages, assumes real significance.

Simulators

One of the rules of the game in chapter 1 was that each player, when asked a question, *must respond*. If the game is played 1,000 times then we collect 1,000 sets of responses, from which the proportion of agreement can be calculated. This corresponds to the demand that every photon which enters a measuring device must either be recorded as having passed or having been absorbed by the polarizer. (In other experiments this corresponds to every electron being found to have spin up or spin down.)

In experimental practice, though, this ideal is never realized. Actual detectors have only limited efficiency. Many, sometimes even most, of the particles simply fail to be detected at all. No usable data can be gleaned from a run of the experiment in which one of the particles avoids detection, so such runs are discarded.

Translating these circumstances back to our game, imagine that each player is accorded a third option: instead of answering either "passed" or "absorbed," one can opt to refuse to answer at all. Data from an experiment in which at least one player remains silent are ignored; only runs in which both respond are counted.

In his 1982 paper "Some Local Models for Correlation Experiments," Arthur Fine devised a clever way to use this third, silent, response to eliminate any need for communication between the two wings of the experiment. (The first such model using considerations about detection was due to Clauser and Horne (1974), but their scheme operates on somewhat different principles.) Fine called these schemes "prism models," but I prefer the term "simulators" for reasons which will soon become apparent. The basic idea is very simple. In our one-bit strategies, one partner depends on information from the other, information as to whether the distant polarizer is set in region 1 or region 0. Suppose no communication can be provided, so even a one-bit message can't be sent. What can the players do?

One thing they can do is simply *guess* into which region the polarizer setting falls. For example, they might guess that the left-hand polarizer *is* set in region 0, and so agree that the right-hand player will use Plan 0. If the left-hand polarizer *is* set in region 0, all is well: both players follow Plan 0 and their behavior is just as if the left-hand setting had been communicated to the right. But if the left-hand polarizer is not set in region 0 then the left-hand player is to *refuse to answer the question at all*. The experiment is deleted from the data on grounds of failure of detection, *and only the cases in which the guess was correct get into the data.*

In this way two players who cannot communicate at all can create the impression that local responses directly depend on the setting of the distant device. Partners who adopt the strategy outlined in the previous paragraph will produce the correlations of figure 6.6 for the runs in which they both answer. The price for this conjuring trick is paid in the coin of efficiency: to simulate sending a one-bit message, 50 percent of the experiments must be disqualified due to one partner refusing to answer. The "detectors" must miss one "particle" in half the runs.

As the informational content of the message increases the efficiency of the corresponding simulation decreases. To simulate communication in the two-bit strategy the partners must guess in which of four equal regions one of the detectors is set, with that partner refusing to answer if the setting is outside the chosen region. Both partners answer, on average, in only one of four experiments and the other three runs are discarded. Simulation of the three-bit strategy requires efficiency to fall to one out of eight, and so on.

We can now calculate a preliminary approximation for the overall efficiency of a mix of strategies which can, without any communication, simulate the quantum correlations. Suppose we want, after a certain number of experiments, to have 1,000 *recorded* experiments which display the correct correlations. As we have seen, when communication is possible, approximately 848 of the 1,000 experiments will use a one-bit strategy. At an efficiency of 50 percent the non-communicating simulators must make twice as many guesses, i.e. they must use 1,696 experiments (on average) to get the 848 correct one-bit responses. To simulate the needed 132 successful two-bit strategies, 528 attempts (on average) are required, the approximately 17 three-bit strategies need $8 \times 17 = 136$ tries, and so on. Taking the sum over all of the strategies, weighted by their respective efficiencies, we find that the simulators need an average of 2,408 experiments to succeed in their illusion. Of these, 1,408 will not be recorded since one of the particles will not be detected. The remaining 1,000 successful experiments will, however, display the $\cos^2 \alpha$ pattern.

Successful simulation of an N-bit strategy requires, on average, 2^N attempts. The significance of our calculation of the average value of 2^N per experiment, 2.408, can now be made clear. On average, a successful simulation of the selected strategy will require 2.408 attempts, and we recover the same result: simulators need, on average, 2,408 tries to achieve 1,000 recorded results.

The simulators can improve their efficiency somewhat by adjustments in their method. We will develop the most efficient scheme in two steps. First, consider the effect of adopting any given θ-strategy. The graph of agreement has the form of perfect agreement when the angle of misalignment between

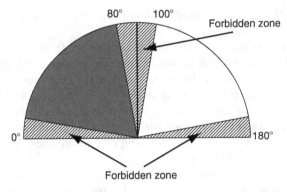

Figure 6.10   Efficient Simulator for θ = 10°

the polarizer is less than the critical angle θ, dropping off linearly to perfect disagreement when the angle lies between 90° − θ and 90°. The difficulty for the simulators is establishing the perfect correlation between 0° and θ. Following the pure zero-bit strategy, the partners will disagree whenever the black/white boundary falls between the two polarizer orientation angles. The likelihood of this occurring is simply proportional to the angle of misalignment.

The simulators circumvent this problem essentially by assigning a *forbidden zone* to one of the particles. The forbidden zone surrounds the black/white borders, and if the polarizer setting for that particle falls within the forbidden zone, the particle refuses to be detected. If we want to assure that the particles never disagree when the angle of misalignment is less than θ, we need to set up a forbidden zone which extends θ on each side of the borders. If the polarizer setting on the chosen side falls outside the forbidden zone, any polarizer setting within θ of it on the opposite side will elicit a matching result.

The method of improving efficiency of our crude one-bit simulators is now obvious. Most of the crude one-bit simulators gratuitously extend the forbidden zone on one side so that the entire zone covers 90° of the possible 180°. More efficient simulators would restrict the zone to cover only 4θ out of the 180°. The more efficient scheme is illustrated in figure 6.10 for θ = 10°. In the regime for 0 < θ ≤ 22.5° the original simulation scheme had a constant efficiency of 50 percent, but for the new method the efficiency is $(\pi - 4\theta)/\pi$ (θ expressed in radians).

What distribution of strategies should these more efficient simulators use? The *observed* frequency of a strategy (i.e. the frequency with which the strategy gives usable data since both partners answer) is the *actual* frequency of the strategy multiplied by its efficiency. We want the observed frequency of every strategy to be proportional to $\rho(\theta) = (\pi - 4\theta)(\cos 2\theta)$. So the actual distribution must be $k(\pi/(\pi - 4\theta))(\pi - 4\theta)(\cos 2\theta) = k\pi \cos 2\theta$, where k is some proportionality constant. Integrating,

$$\int_0^{\pi/4} k\pi \, \cos(2\theta)\,d\theta = k\frac{\pi}{2} = 1$$

we find that $k = 2 / \pi \cong 0.6366$. The overall efficiency for this more efficient simulation is about 63.66 percent: on average, 1,571 experiments will yield 1,000 usable results and the results will show the proper pattern of correlation.

In complete detail, then, one way to achieve the quantum correlations without any communication is as follows. While still in the room, you and your partner randomly choose an angle θ between 0° and 180° (using the constant probability density $1/\pi$). You construct an instruction set by coloring the region from ϕ to $\phi + 90°$ (mod 180) black. Next, an angle θ between 0° and 45° is chosen using the probability density $2\cos (2\theta)$. The regions which extend θ on either side of the black/white boundaries on one of the instruction sets are designated the forbidden zone. If the partner with that instruction set is asked a question that lies within the forbidden zone (if the polarizer is oriented at one of the forbidden angles), he refuses to answer. In the long run you will both answer 63.66 percent of the time, and in the long run the statistics displayed for the cases in which you both answer will match the target curve: you will seem to be doing what the photons do. Let us call this plan of action *Scheme A*.

Scheme A can be improved on slightly by dividing the forbidden zone evenly between the two particles rather than concentrating it in one of the instruction sets. Again, we want the θ-strategy to guarantee agreement between recorded pairs of particles so long as the angle between the polarizers is less than θ. Disagreement between responses only occurs when the polarizer settings fall on opposite sides of a black/white border in the instruction set. For this to occur when the angle between the polarizers is θ requires that *one* of the polarizer settings falls within $\theta/2$ of the border. So by adding to *each* instruction set a forbidden zone which extends $\theta/2$ from each black/white boundary the θ-strategy can be simulated. Call this method of simulation *Scheme B*.

Scheme B poses a slightly greater analytical problem than Scheme A. Complications arise because the efficiency of a Scheme B strategy varies with the angle of misalignment between the polarizers. When the polarizers are perfectly aligned the probability of getting both particles to respond is $(\pi - 2\theta)/\pi$, the probability that the chosen orientation of the polarizers falls within the common forbidden zone around one of the two boundaries. This efficiency decreases linearly as the angle of misalignment grows, reaching $(\pi - 4\theta)/\pi$ when the angle is θ. The efficiency remains at this level up to the maximum angle of misalignment, $\pi/2$.

To spare the reader tedious calculational detail, we will simply quote the main results.[3] In order for the recorded pairs to show the desired $\cos^2 \alpha$ correlations, Scheme B strategies must be chosen in accord with the probability density

$$\rho\,(\theta) = \frac{1}{\cos^4 \theta}\left(\left(\frac{\pi}{2}+1-2\theta\right)(1+2\sin^2\theta)-4\sin\theta\,\cos\theta\right).$$

Using this weighting, Scheme B achieves the desired results with an efficiency of $(\pi + 2 - 4\alpha)/2\pi \cos^2 \alpha$, where α is the angle of misalignment measured in radians, when α is between 0° and 45°. When the polarizers are perfectly aligned, both particles will respond about 81.8 percent of the time. Efficiency decreases until, at the misalignment of $\pi/4$ (45°), it reaches a minimum of 63.33 percent, equal to Scheme A, where it stays from $\alpha = 45°$ to $\alpha = 90°$.

There is no obvious way to improve on the efficiency of Scheme B. The total area of the forbidden zones cannot be decreased and Scheme B constitutes the best distribution of the forbidden zones between the two instruction sets. I conjecture that Scheme B is the best one can do in the absence of any information-transmitting connection between the two wings of the experiment. One can get the right correlations between the particles in those cases when both particles are detected, but joint detection of the particles must fail from 18.2 percent to 36.3 percent of the time.[4]

Does Nature Simulate?

The possibility of systems which instantiate Scheme A or Scheme B raises a doubt about our fundamental thesis: viz., the existence of some real superluminal connection in nature. Simulators require no direct connection of any kind between the two sides of the apparatus. So we must ask two questions. First, how could we determine whether a simulating strategy is being used? And second, how intrinsically plausible is it that simulation accounts for the phenomena already observed in the laboratory?

The most obvious mark of simulation is the ineliminable inefficiency in the rate of particle detection. No matter how the design of polarizers and photomultiplier tubes improves, a considerable fraction of the particles must escape detection. In particular, when the polarizers are 45° or more apart no more than 63.66 percent of the experimental runs will give results where both photons are detected. This result would itself constitute a problem for the quantum theory, since it implies no such limit on detector efficiency.

None of the experiments done to date employs detectors efficient enough to rule out simulation. T. K. Lo and Abner Shimony (1981) have proposed an experiment which should be able, using existing techniques, to achieve high enough efficiency to rule out simulators. But even before such high efficiencies have been achieved, simulation would make itself known. For it is not simply that simulators demand inefficiency, but also that the inefficiency takes an unusual form. Normal inefficiencies in experimental set-ups are due to limitations on the ability of detectors to capture photons or of polarizers to function ideally. Such failures ought to occur on each side of the experiment independently of what happens on the other. That is, for normal inefficiency there should be a number, Q, which measures the probability of a photon which reaches an analyzer being detected, and Q ought to characterize each detector independently of the other. In this normal case, both photons will be detected on a proportion Q^2 of the runs, *one* photon will be detected $2(Q - Q^2)$ of the time, and *neither* will be detected $1 - 2Q + Q^2$ of the time. These proportions should hold irrespective of the angle of misalignment between the polarizers.

Simulator inefficiencies do not fit this pattern. In Scheme A it is *never* the case that both photons fail to register since the forbidden zone is concentrated in one instruction set. Even when overlaid with normal inefficiency this extra proportion of experiments where only one photon shows up would be noticeable. Scheme B gives rise to different sorts of anomalies. Most obviously, as already noted, in Scheme B the proportion of usable cases, runs in which both photons are detected, varies with the angle between the polarizers. This has already been ruled out by experimental observation. Furthermore, when the polarizers are aligned, the "built-in" inefficiency on each side is correlated with the inefficiency on the other: either both photons are detected or neither are. These peculiarities would also be noticeable even when normal detector inefficiencies are added to each side of the experiment. Hence analysis of correlations among inefficiencies in the two detectors could reveal the traces of Schemes A and B even while the net inefficiency remains high.

These observations are unlikely to send experimentalists scurrying to analyze detector inefficiencies. Scheme A and Scheme B do provide examples of how quantum mechanical correlations could appear in experiments without the existence of any superluminal influences, at the price of ineliminable inefficiency in the detectors. But both schemes suffer from defects which fall under the general rubric "*ad hoc.*" The models were constructed for the sole purpose of replicating predictions which arise in a very natural way from quantum mechanical principles, but the models are not based on any physical principles which recommend themselves on independent grounds. They do not look very "natural" – the probability density $\rho(\theta)$ for Scheme B being notably baroque. Furthermore, they rely on fundamental mechanisms which directly oppose the quantum mechanical picture.

In the standard interpretation of quantum mechanics with wave collapse, every particle pair is created in exactly the same physical state. Each pair is described by the same wave-function. This wave-function is, furthermore, perfectly symmetrical under rotation about the axis of motion of the particles: no direction in space is singled out. Wave collapse provides a means by which the state of the particle on one wing is altered by measurements made on the other, altered in a way which reflects the orientation of the distant polarizer. Wave collapse transmits an infinite quantity of information about the state of the distant apparatus, and this information transmission connects the two photons in a way which creates violations of Bell's inequality.

In Bohm's theory it is no longer true that all photon pairs are created equal: differences in photon position at creation ultimately account for the different results in successive runs of the experiment. But even so, in Bohm's theory the *wave-function* of each particle pair is the same, just as it is in standard quantum theory. That wave-function always evolves in accord with Schrödinger's equation, and its evolution is affected by the setting of each polarizer in such a way that a change in the distant polarizer setting can, in some cases, cause a change in the local response. A photon can thereby be influenced by the orientation of the distant polarizer.

In Schemes A and B, though, nothing even vaguely resembling these pictures remains. Every photon pair is created in a state which is *not* symmetric under axial rotation: the orientation of the black/white borders in the instruction set picks out preferred directions. The choice of the size of the forbidden zone adds another difference among the underlying states. The surface similarity between the simulators and quantum mechanics goes no deeper than the surface: both generate the same statistics for observed photon pairs, but they do so by entirely different means. The statistics produced by the simulators are determined entirely by the choice of $\rho(\theta)$, and our choice was guided by no other consideration than mimicking the quantum predictions.

For all these reasons simulator schemes are unnatural and ugly. No physicist would take them seriously as proposals in competition with quantum theory, and anyone would be irrational to suppose that scrutiny of inefficiencies in Aspect's experiments would yield any evidence of simulation. So why have we gone to the considerable trouble of constructing these models?

The answer is twofold. First, although the models are *ad hoc*, ugly and unnatural, they represent a possibility for explaining the experimental results we have at present without superluminal connections of any sort. This loophole can in principle be closed by further experiments with more efficient detectors, but it has not been closed yet. Positions in logical space deserve attention even if they have nothing else to recommend them.

More important than this, though, is the *way* the simulators were derived, the principles behind their construction. We began by seeing how close one can get to the quantum mechanical correlations if no information about distant polarizer settings is available. The quantum mechanical correlations cannot be reproduced (if detectors are 100 percent efficient). We went on to ask what minimum amount of information about distant settings would allow reproduction of the quantum correlations. The simulator models arose from considerations of how inefficiency could be used to mimic information transmission.

The key to the whole analysis is the recognition that, in the absence of inefficiency, *Bell's inequalities can reliably be violated only when the response of one of the particles depends (at least sometimes) on the question asked its partner.* This observation reinforces and justifies the results of the last chapter: violation of the inequalities implies superluminal causal connections. It also points up the fact that dependence on the distant *polarizer setting* is crucial. Jarrett's division of theories into those that violate outcome independence and those that violate parameter independence is again seen to be misleading: *any* successful theory must postulate some influence of a distant "parameter" (i.e. the polarizer angle) on the response of a local photon. Without such dependence the quantum statistics cannot be recovered.

These results provide an antidote to one strain of argument which occasionally arises in the literature on this problem. It is sometimes claimed that the violation of Bell's inequality ought not to be puzzling to us at all. Our sense that such violations are surprising, or that they reveal something deep about locality in the physical world, is attributed to our use of outdated intuitions or mistaken inferences. The central thrust of these claims, exemplified by Fine (1989a), van Fraassen (1982), and Redhead (1987, p. 95), is that violations of the inequality produce in us a sense of surprise only because we have illegitimately assumed some standard of explanation which is rooted in classical physics, and more particularly in the assumption of determinism. These unexamined assumptions lead us to demand more in the way of explanation than can legitimately be demanded of a stochastic theory; they are the products of some latent philosophical prejudices or incorrect modal inferences.

But what is puzzling about the quantum statistics for pairs of photons is simply that they cannot be reproduced by *any* means, stochastic or deterministic, if there is no information transfer between the wings of the experiment. Those who doubt this claim are challenged to reproduce the statistics without such transfer, to play our game with one partner not allowed to communicate with the other. They may use whatever stochastic devices they like: decaying uranium atoms, flipped coins, and so on. They may use whatever modal logic appeals to them. They may invoke whatever standards of explanation they please. The bottom line is: can you get the *numbers* without having information about the setting of the distant polarizer available to one partner? The answer is that you cannot. This dependence on distant events does not violate any classical physical principles, if "classical" is taken in the usual way. It does at least seem to conflict with relativistic constraints. So we have had reason to be worried about our photons all along.

Schemes A and B allow us to begin to quantify the minimum amount of information transfer needed to produce the correlations (roughly, the efficiency of each model is inversely proportional to the average value of 2^N, where N is the number of bits of information transmitted). They also refute a diagnosis offered by Arthur Fine concerning the fundamental principles which underlie the derivation of Bell's theorem. Fine's 1982 paper discusses some rather crude simulator models, the so-called "prism" models. These models introduced the principles which drive Schemes A and B: each particle will respond only to a certain range of possible polarizer settings. Fine only deals with the highly restricted situation in which each polarizer can be set in only *two* positions, not the continuum of possible orientations of which we have treated.[5] In his "minimal model" (Fine 1982, p. 284) each particle will respond to only one possible setting. In essence, each particle guesses exactly how its polarizer has been set and refuses to answer if the guess is wrong. The efficiency of this minimal model is 0.25 and would evidently drop to zero if expanded to cover all possible polarizer settings. (Your chance of guessing correctly the exact orientation of the polarizer in this general case is zero.)

Fine knows that he can do better than the minimal model, and so goes on to propose a "maximal model." In this model the particles only guess what one of the two polarizer settings is *not*. That is, of the four possible individual analyzer settings (two on each side) there is always exactly one to which one of the photons will not respond. Fine divines that this is "the best one can do," hence the name "maximal." The reasoning is as follows:

> The Wigner version of the Bell-inequalities makes plain that the deviation from quantum mechanics comes from the assumption that all the response functions [i.e. functions which determine how a particle will respond to a measurement] are defined for all λ [i.e. for all the hidden states]. For this leads to there being well-defined multiple distributions, like $P_{A_1 A_2 B_1 B_2}$ which are never well-defined in quantum theory itself and from which the Bell inequality follows immediately ... It can happen, however, that distributions like $P_{A_1 A_2 B_1}$ *are* well-defined quantum mechanically, even though A_1 and A_2 are incompatible. For instance, if A_1 were strictly correlated with B_1, then one would have that $P_{A_1 A_2 B_1} = P_{A_2 B_1}$. Hence the best one can have, without presupposing something in direct conflict with quantum mechanics, is that three of these four response functions may be defined for certain measurable sets of λ (but not all four at once). This is the guiding idea of the *maximal model*. (Fine 1982, pp. 286–7)

Fine's argument, supposedly derived from reflections of the role joint distributions play in Bell's theorem, is meant to establish that any model not in direct conflict with quantum mechanics will postulate hidden states in which each pair of photons will respond *at most* to three of the four possible measurements which could be made. This would result in a maximal efficiency of 0.5. (Two measurements are made per experiment, one on each side, and there is a 50 percent chance that the forbidden measurement will be chosen.) Furthermore, the argument implies that the hidden state can be constructed to respond to three possible measurements only when two of the measurements are in the same direction (on different wings).

Schemes A and B show all of Fine's assertions to be false. First, both models achieve efficiencies higher than 50 percent for all possible polarizer settings. Indeed, Scheme B can produce violations of Bell's inequality with efficiencies higher than 80 percent. (Let the two polarizer settings on one side be 0 and 0.01 radians, the settings on the other be 0.01 and 0.02 radians.) So Fine's supposedly maximal model can be greatly improved upon. But more critically, the entire line of argumentation is shown to collapse. According to Fine, no measurable set of underlying states can be such as to respond to all four possible observations. But as is obvious, many states in both schemes will respond to all four possible measurements no matter which four you choose. Indeed, in the case mentioned above, with over 80 percent efficiency, *most* pairs of photons would respond to all four of the settings.

Even more striking, the strategies used in Scheme A and Scheme B are not constrained at all in the way Fine conjectures. *Every* θ-strategy has a well-defined response function (pass or absorb) for an *infinite* number of possible measurements, including an infinity of joint observables which are quantum-mechanically incompatible with one another.[6] In fact, if Fine's argument that at most *three* observables can be simultaneously well-defined were valid, the efficiency in the general case, with a continuum of possible polarizer settings, would drop to zero, as W. D. Sharp and N. Shanks show (1985, p. 555).

Where did Fine go wrong? His general argument starts from derivations of Bell's inequality which rely on the following assumption: for every physical state the particles could be in, the probability for getting a particular response (passage or absorption) for a given observation is well defined *and the probability for getting a pair of responses is just the product of the probabilities for the individual responses* (i.e. the hidden states screen off the behavior of each particle from that of the other). Since no model which obeys this constraint can violate Bell's inequality, Fine starts thinking of ways to restrict the states on which joint distributions are defined by restricting the response functions.

But the most obvious cases where the assumption stated above breaks down is when the two events are causally connected. If my partner and I both flip fair coins the probability that each of us will get heads is 0.5 and the probability that both of us will get heads is 0.5 × 0.5 = 0.25. But if my partner flips a fair coin and communicates the result to me, the individual probabilities that we will each get heads remains 0.5 and the joint probability of both getting heads is also 0.5. Under conditions such as these, violations of Bell's inequality are trivial to achieve.

By thinking about communication, and simulation of communication, we have been able to beat Fine's "maximal" limit even while constructing models that agree with quantum mechanics for the full range of possible observations, not just a highly restricted set. This suggests that information transfer, the causal dependence of one side on the distant analyzer setting, strikes closer to the heart of the situation than does reflection on incompatible quantum mechanical operators. Indeed, I will stick my neck out as much as Fine did: I believe that Scheme B is the most efficient means of mimicking the quantum mechanical correlations if there are no causal connections between the wings of the experiment. Detectors which achieve efficiencies above 82 percent for joint observations (or above 64 percent when the polarizers are misaligned by 45° or more) would be inconsistent with *any* simulator model.

In the last chapter we saw that reliable violations of Bell's inequality imply a causal connection between the two sides of the experimental apparatus. In the quantum mechanical picture which includes wave collapse the influence of a measurement made on one side on the state of the other photon is massive: the photon's physical state is changed to one which reflects the setting of the distant polarizer and the response of the distant photon to the measurement. The information conveyed by this change in state is infinite since the new physical state is compatible with only two out of a continuum of possible distant settings (viz., the setting parallel to the photon's polarization and the setting perpendicular to that direction). If Nature works as this model suggests then she is quite profligate with superluminal information transmission.

We have seen that efforts to reduce the needed amount of information can be quite successful, reducing the average information transmitted per experiment to less than one bit. But that residual superluminal connection is ineliminable.

Does superluminal information transmission automatically violate relativity theory? As we have already seen, it need not give rise to causal paradox, if the information is not available for general use. Nor need it imply the existence of preferred inertial frames: even superluminal signals could avoid this pitfall. But these observations should not entirely allay our fears. We do not have a relativistically invariant version of theories such as Bohm's that avoid wave collapse altogether. The possibility of a completely relativistic theory of wave collapse is taken up in chapter 10.

Some superluminal connections exist in nature. The question which now faces us is whether those connections are consistent with the relativistic account of space-time structure. In the case of Special Relativity, such consistency can only be demonstrated by displaying a Lorentz invariant theory which will account for the quantum phenomena. The quest for such a theory brings us into closer contact with the quantum theory itself than we have been heretofore, and will plunge us deeper into the mists of speculation.

Notes

1 Any target probability function $f(\theta)$ defined for $0 \le \theta \le \pi/2$ such that

1 $f(0) = 1$

2 $\dfrac{\partial^2}{\partial\theta^2} f(\theta) \le 0$ for $0 \le \theta \le \pi/4$

3 $f(\theta) = 1 - f(\pi/2 - \theta)$

can be matched by this method. The appropriate weighting function is

$$\rho(\theta) = -\frac{1}{2}(\pi - 4\theta)\frac{\partial^2}{\partial\theta^2} f(\theta).$$

If all strategies from 0 to $\pi/4$ are available, the match will be exact. If only strategies from 0 to $\phi < \pi/4$ can be used then one uses the weighting function above for the available strategies $< \phi$ and concentrates the remaining weight on strategy ϕ. The resultant curve will match the target exactly from 0 to ϕ with a linear connecting piece from ϕ to $\pi/2 - \phi$.

2 The two-bit strategy we have chosen allows the recipient of the message to place the distant setting somewhere within a disconnected set of regions which together cover one-fourth of the total possibilities. Other strategies are possible, strategies which allow one to locate the distant setting within a more connected region. Some two-bit strategies have features which allow a perfect match to the target curve from 0° to 45° with deviations in the region from 45° to 67.5°. No exhaustive analysis of all possible strategies exists, but I see no reason to think that some other scheme could do better than this.

3 In general, any function $f(\phi)$ which satisfies the three criteria listed in footnote 1 can be simulated by Scheme B. The appropriate weighting function is the solution of

$$\rho(\theta) = \frac{\partial^2}{\partial\theta^2}\left(\frac{\frac{\pi}{2}(1 - \int_0^{\pi/4} \psi\rho(\psi)\,d\psi) - \theta}{f(\theta)} \right).$$

Both this equation and that for $\rho(\theta)$ correct errors in my 1992 paper, although all of the substantial claims about the models made in that paper are correct.

4 It is obvious that the behavior of particles in the GHZ scheme can be simulated with a net efficiency of 50 percent – if one photon refuses to answer for one setting, the other five slots in figure A.1 can be filled in with numbers that satisfy the constraints. I have no idea what would happen if one allowed each of the three analyzers in the GHZ scheme to take on the full range of settings.

5 Abner Shimony (1989, p. 31, n. 8) challenges Fine to demonstrate that the prism models can be expanded to cover all possible polarizer settings; Schemes A and B answer this challenge.

6 In terms of Fine's later (1989b) paper, Schemes A and B solve the $\infty \times \infty$ problem (i.e. an infinite set of possible observations can be made on each side), and obviate the need for a "rigging" of the models to the experimental set-up. These models beat the efficiency constraints announced in that paper since the probability of detecting neither particle is not the square of the probability of detecting a single particle.

7

Points of View

Relativity is commonly taken to imply that the speed of light functions in nature as some sort of absolute limit. In the previous four chapters we have examined the claims that it is a limit on the velocity of energy transmission, of signal speed, of causal connection and of information transmission. We have found that reliable violations of Bell's inequality do not entail transgression of the first two restrictions, but do require violation of the last two. And we have further found that none of the restrictions can be derived, in any strictly formal sense, from the Lorentz transformations or from the fundamental relativistic space-time structure. On the contrary, explicitly relativistic theories of tachyons and of superluminal signals have been constructed. The fundamental feature of the Lorentz transformations is that they leave the speed of light invariant, not that they render it an insuperable boundary. So if Bell's theorem, together with the results of Aspect's experiments, demonstrates that the world is non-local, we must turn our attention to the question of the compatibility of that non-locality with the relativistic picture of space-time. We must take up the question of whether a complete quantum theory can be rendered Lorentz invariant.

We will not immediately attack this problem head on. Instead, we will return for a moment to the classical realm, and to the classical analog of Lorentz invariance, namely Galilean invariance. The reason for this is twofold. First, Galilean invariance is a simpler and more tractable formal property, and so provides a good training ground to develop some of the technical machinery. But second, and perhaps more important, Galilean invariance does not display some of the fundamental properties of Lorentz invariance. So intuitions honed on the Galilean case, as the intuitions of most students

Quantum Non-Locality and Relativity: Metaphysical Intimations of Modern Physics, Third Edition. Tim Maudlin.
© 2011 Tim Maudlin. Published 2011 by Blackwell Publishing Ltd.

of physics are, can be quite deceptive if transferred to the arena of Special Relativity. It will be easier to point out these disanalogies, and so to avoid them, if we have studied the classical case first.

Galilean Transformations and Galilean Invariants

The inertial frames of classical physics are coordinate systems with two properties. First, at any given moment, space is coordinatized by a rectangular Cartesian grid. Second, each of the spatial coordinates on the grid follows an inertial (i.e. straight) trajectory through the space-time. (It is really only this second property which makes the frame inertial.) Suppose we have two such frames, primed and unprimed, which exactly coincide at $t = t' = 0$, such that the primed frame is moving at a constant speed v with respect to the unprimed frame. For simplicity, we imagine that the unprimed frame is moving in the positive x-direction. Then the equations which give the location of a point in the primed coordinates in terms of its location in the unprimed coordinates are:

$$t' = t$$
$$x' = x - vt$$
$$y' = y$$
$$z' = z.$$

This is a Galilean transformation.

If a physical theory respects the Galilean transformations then it will take the same form in all inertial coordinate systems. Physically, this means that none of the inertial frames is preferred over any other, so there is no "absolute space" or "absolute velocity." Different coordinate systems will attribute different velocities to a particle but none will be any more "correct" or "true" than any other.

It is instructive to consider how a classical theory might achieve Galilean invariance. The most obvious route is through quantities which are themselves invariant under the Galilean transformations. For example, although velocities are changed under such a transformation, acceleration (change in velocity) is not, for if we add the same constant term to two velocities their difference remains unchanged. Newtonian gravitational theory achieves Galilean invariance by this means. The fundamental dynamical law of Newtonian mechanics is $F = ma$. The acceleration a, as we have seen, is Galilean invariant. So too is the inertial mass m in Newtonian theory. So the necessary and sufficient condition for any Newtonian theory to be Galilean invariant is that the force F be the same in all inertial frames. The Newtonian gravitational force $G = GmM/r^2$ satisfies this constraint since the masses, the gravitational constant G and the instantaneous distance r between the two masses are all invariant quantities. So if a system satisfies the Newtonian gravitational laws in one inertial frame it will satisfy it in all of them. (Since distance between non-simultaneous events is not Galilean invariant, it will take some work to produce a Newtonian theory of gravity without instantaneous action at a distance!)

Because different inertial frames disagree on the velocities attributed to particles it may seem that no laws in Galilean space-time can make reference to velocities. But there are several ways around this apparent constraint. That the speed of sound is about 760 m.p.h., for example, can perfectly well be explained in a Newtonian context. The relevant speed is not the speed of the sound wave as calculated in each reference frame but the speed of sound *relative to the medium*. The mass of air in which the sound propagates picks out a preferred frame of reference, namely the frame in which it is at rest. Only in this frame does the sound travel isotropically.

The appearance of velocities (and the implied value of the speed of light) in Maxwell's equations of electro-magnetism was originally interpreted as implying the existence of a material medium for the electric and magnetic field, the so-called lumeniferous ether. Attempts to detect the frame of reference in which the ether is at rest failed, providing, at least in retrospect, justification for Special Relativity. But what would have happened if the ether frame had been detected, if light propagated isotropically in only one inertial frame?

Opinion on this issue is divided. On the one hand, one can argue that the existence of such a naturally preferred frame of reference would vindicate Newton's theory of Absolute Motion. The velocities appearing in Maxwell's equations would be interpreted as Absolute Velocities, i.e. velocities with respect to Absolute Space. It would then be, as Newton believed, nothing more than an unfortunate coincidence that the laws of Newtonian mechanics do not allow for the experimental identification of the state of Absolute Rest. Space-time would have the structure of parts of Absolute Space persisting through time, thereby picking out the preferred inertial frame.

On the other hand, one can argue that the discovery of such an optically preferred frame would no more imply Absolute Space than the existence of an acoustically preferred frame does (see, for example, Sklar 1974, p. 197). According to this view, the ether frame would be just that: the inertial frame picked out by the state of the motion of the ether. Whether the ether itself was at Absolute Rest or in Absolute Motion could not be decided by any optical experiment, and, indeed, the meaningfulness of the very question could be denied. The ether could exist in a Galilean space-time in which no Absolute Velocities are defined.

Since we will be considering parallel questions of which sorts of phenomena are compatible with Minkowski space-time structure, it behooves us to take a stand on this case. But our ultimate verdict will depend on further details of the nature of the new discovery.

If the ether picks out different inertial frames at different places, as various masses of air pick out different "rest" frames, then the discovery could not be taken to vindicate Absolute Space. Newtonian Absolute Space is a single inertial frame for the entire universe. The so-called "ether drag" theories in which the Earth entrained its ether would give no support to the rejection of Galilean space-time.

Even if the ether everywhere defined the same frame, some further questions would be in order. Does the ether have the sorts of properties, such as mass, which are associated with material substance? Can the ether be thinned out or compressed? If so, then it is better interpreted as a material occupant of space-time rather than an intrinsic feature of space-time itself.

Finally, if the ether had none of these properties, if it were sufficiently "ethereal" to escape classification as a material substance, then the existence of the naturally preferred frame of reference would have to be taken as evidence that Newton was right and that there is more to space and time than the Galilean structure. After all, it is hard to see how one could argue *a priori* that Newton's conception could not be right, and difficult to imagine what evidence for that position could be stronger than the occurrence of velocity terms in fundamental laws of nature, velocity terms which did not seem to refer to velocities relative to any material.

The symmetries of space-time are the symmetries of the vacuum and the vacuum is the absence of material fields. If no natural phenomena pick out a preferred reference frame in a vacuum one ought not to posit that one reference frame has a special status. This rule of thumb is not an algorithm since there may always be arguments about what constitutes a vacuum. The die-hard Galilean could always insist that the natural phenomena, such as the propagation of light in the absence of all known material fields, still depend on the presence of some highly attenuated substance in the space. But if the existence of such a substance cannot be independently observed then the argument becomes progressively weaker, until the most acceptable response is the admission of more structure to space-time itself.

Finally, one further possibility must be examined. The laws of nature might be given in terms of velocities rather than accelerations, and the velocities might not be referred to some rest frame of a medium, and yet the laws could hold in all inertial frames. Non-relativistic Bohmian mechanics provides an example of this situation. As we saw in appendix B, Bohm's theory posits that the quantum wave-function determines a velocity field for the particles. How can Bohm's theory then avoid the existence of a preferred reference frame, namely, the frame in which the particle velocities are those given by the wave-function?

The solution lies in the observation that the wave-function itself *is not invariant under a Galilean transformation*. A Galilean boost to a frame moving with velocity v_0 relative to a given frame is implemented by the transformation on the wave-function:

$$\psi \rightarrow \exp\left(i\frac{m}{\hbar}v_0 \cdot q\right)\psi.$$

(Durr, Goldstein, and Zanghì 1992, section 3).

The transformed wave-function, when plugged into the laws of motion, generates the appropriate velocities for that reference frame. Thus a first-order Galilean theory can exist even in the absence of any material medium which can pick out a preferred reference frame.[1]

Collecting together all of these threads, physical theories appropriate to a Galilean space-time should have three features. First, the laws of the theory (expressed in terms of coordinates) should hold in all inertial frames. This could be because (as in the Newtonian case) the laws are framed in terms of invariant quantities or (as in Bohm's theory) because the quantities transform in such a way under the Galilean transformation that the equations continue to hold.[2]

Second, one wants the physical state descriptions of systems to transform under the Galilean boosts in a manner consistent with the universal validity of the laws. That is, if we start with a system at time 0 described in unprimed coordinates then it shouldn't matter whether we first use the laws to generate the state at time t and then transform to the primed frame or first transform to the primed frame and then use the (invariant) laws to generate the state at t. In technical terms, this amounts to saying that the diagram

$$S_x(0) \rightarrow S_{x'}(0)$$
$$\downarrow \qquad \downarrow$$
$$S_x(t) \rightarrow S_{x'}(t)$$

is commutative, where the arrows pointing to the right represent Galilean transformations and the downward arrows represent time evolution according to the laws. $S_x(t)$ should turn out the same whichever path we take to get there.

Finally, one wants the vacuum state to be Galilean invariant, i.e. one wants the vacuum state to have the same functional form when expressed in any inertial coordinates. If the vacuum state picks out one frame of reference over another then one must ascribe that non-Galilean structure to space-time itself.

As the case of the lumeniferous ether illustrates, it is not always clear how one determines when these criteria have been met. Any anisotropy in observed phenomena could be attributed to the presence of a hidden material medium if one is desperately anxious to preserve the Galilean structure. But these sorts of criteria, applied with good taste, are the best we have to go on when trying to determine the intrinsic structure of space-time. As we will see, the story becomes somewhat more complicated when we shift from Galilean to Minkowski space-time.

A Brief Preliminary: Why Worry?

Devoting so much space to the examination of Lorentz invariant formulations of quantum theory may appear as a waste of time to many physicists. After all, why not just look it up in a standard reference on relativistic quantum field theory? Quantum field theory is an old and well-established discipline, and can be motivated exactly by relativistic considerations. Why is not the existence of relativistic quantum field theory a direct proof by construction that quantum theory and Relativity are compatible?

In quantum field theory the analog of Schrödinger's equation is relativistically invariant, and the quantum states transform properly under the Lorentz transformation, but these features could secure the credentials only of a theory *which has no wave collapse*. In any orthodox theory the wavefunction is complete and hence must collapse, so we must consider whether collapses could be generated in a relativistically invariant way. We turn to this question presently. It is sometimes claimed that the relativistic credentials of relativistic quantum field theory are secured by the so-called Equal Time Commutation Relations (ETCRs). These state that operators which represent field quantities at space-like separation all commute. This is supposed to show that the theory contains no superluminal action. But all of our preceding investigations reveal the emptiness of this claim. We have seen that reliable reproduction of the quantum statistics demands superluminal causation and superluminal information transmission. And this has been proven despite the fact that the relevant quantum mechanical operators, viz. the operators for the polarization of the two photons, commute with each other. Commutation of operators for two space-time regions does not guarantee that causal connections do not exist, or indeed are not *required* to exist, between the regions.

The fact that the operators commute does show that measurements made on one side do not affect the long-term statistics on the other, i.e. that no signals can be sent from one region to the other by manipulating polarizers on one side. We acknowledged this point in chapter 4, but still have found that superluminal influences are required to explain the correlations between the two regions. So citing the ETCRs shows nothing about the need for superluminal causation in quantum field theory.

The reason that quantum field theory may seem to be evidently compatible with relativity is that the non-local influences in orthodox quantum theory are carried by wave collapse, and wave collapse is commonly ignored in physics texts. But since we have been adamantly sweeping the dirt out from under rugs, we must now face squarely the question of how various interpretations of quantum theory, those with and without collapse of the wavefunction, can fare in the relativistic domain.

Lorentz Invariance: Collapse Theories

The Lorentz transformation is a more complicated business than the Galilean transformation. In particular, shifting from one Lorentz inertial frame to another entails changes not only in the velocities ascribed to objects but also in which events are taken to be simultaneous. Since the property of being simultaneous is not Lorentz invariant, no invariant laws can make use of it.

Relativistic theories that involve causal processes that propagate at or below the speed of light generally have no need of the notion of simultaneity. Most obviously, electromagnetic theory can dispense with it quite neatly since the light cones, along which electromagnetic forces have their influence, are Lorentz invariant surfaces. Part of the difficulty involved in formulating a relativistically acceptable theory of gravitation arose from the fact that the Newtonian theory did postulate instantaneous action-at-a-distance. (Ironically, the problem arises not because the action is unmediated, the feature to which the mechanical philosophers objected, but because it is instantaneous. As we have seen, the instantaneousness of the Newtonian force helped make the theory Galilean invariant.) Overcoming these difficulties necessitated positing a delay in the gravitational influence, and was only successfully completed with the General Theory of Relativity.

The classical formulations of non-relativistic quantum theory also posit an instantaneous non-local change: wave collapse. When our two photons leave the source each is in a state of indefinite polarization (or each is not in any state of polarization). At the moment that the first photon is observed the wave-function describing the pair undergoes a sudden change such that the unobserved partner assumes a definite state of polarization which matches its partner. In this way the perfect correlations of the EPR experiment are maintained (and the correlations which violate Bell's inequality generated).

In Minkowski space-time this theory of wave collapse no longer makes sense. The collapse can be instantaneous in at most one reference frame, leading to two possibilities: either some feature of the situation picks out a preferred reference frame, with respect to which the collapse is instantaneous, or the collapse is not instantaneous at all.

The analogy with gravitational theory might suggest the latter course as the most obvious: why not build a delay into the account of wave collapse? Why not, as in General Relativity, have the collapse propagate along the future light cone of the measurement event, thereby allowing a Lorentz invariant description?

The obvious answer is that such a delayed collapse would come too late. Since polarization measurements can be made at space-like separation, and since the results on one side must be influenced by the collapse initiated at the other, delayed collapses won't work. A more promising possibility would be to use the *back* light cone of the measurement as the locus of the collapse. Again, the locus is manifestly Lorentz invariant, and now the collapse could carry the requisite information to the right place on time. Let's look into this suggestion a bit more closely.

The first thing to note is that since the photons have traveled from the source to the detector along a light-like path, if the wave collapse takes place along the past light cone of the measurement then *the photons are always in the collapsed state*. It is therefore rather misleading to call this a theory of wave collapse at all: no collapse occurs since no uncollapsed state is ever produced. This criticism does not invalidate the suggested theory, but does show that it may be more perspicuously presented by omitting reference to any collapse. Once this is done the essentials of the theory come clearly to light.

The essence of the theory is that a measurement of a photon can influence events which lie on the back light cone of the measurement. Since the production of the photon is one such event, the nature of the photon produced can be in part determined by the type of measurement later carried out on it. In our canonical photon-pair experiment, this means that the settings of the detectors can influence the state of the pair of photons produced.

It is immediately obvious that such a theory can in principle account for violations of Bell's inequality, or indeed for any correlations which could be imagined. It is as if in our imaginary game one were told the questions to be asked *before* ever leaving the room. It would be no great trick for the partners to arrange their responses to display any specified statistics.

We need not artificially construct a theory along these lines since proposals with such backwards causal influences have already been developed. We will take as a clinical example the *transactional interpretation* of quantum mechanics due to John Cramer, which postulates that quantum events are determined by the interaction of wave-functions that propagate both forward and backward in time (Cramer 1980, 1986). Taking as his inspiration the electrodynamic theory of Wheeler and Feynman, which employs both advanced and retarded electromagnetic radiation (Wheeler and Feynman 1945), Cramer postulates that not only does an emitter send a wave-function forward in time to an absorber, but the absorber sends one backward in time to the emitter. The state of the absorber (for example, the orientation of a polarizer) thereby influences the emission event, allowing the production only of photons in appropriate definite states of polarization. Cramer uses this mechanism to explain how the quantum correlations can be produced at space-like separation (1980, pp. 368–70; 1986, pp. 667ff.).

The price to be paid in any such theory is the acceptance of explicit backwards causation, effects which precede their causes in every frame of reference. Producing a consistent theory which incorporates both directions of causation is a rather tricky business, and is especially difficult if the theory is to be *stochastic* rather than deterministic. The problem is that the stochastic outcomes at a particular point in time may influence the future, but that future itself is supposed to play a role in producing the outcomes. As an illustration of the complexities this generates, let's consider Cramer's theory in more detail.

Cramer explicates his theory with the use of an admittedly fictional device he calls a "pseudotime sequence" (1986, p. 661). Suppose we have a radioactive atom which may decay, sending a beta particle[3] either to the right or to the left, and suppose there are absorbers set up on the right and left. In pseudotime, the interaction which occurs may be described as follows. The atom sends out an "offer" wave forward in time. This wave is the standard, uncollapsed wave function. *Both* of the absorbers receive this offer wave at a time later than the emission, and each sends back in time a "confirmation" wave which is the time reverse of the offer wave it received. These two confirmation waves arrive at the emitter (in particularly nice mathematical form: they are equal to $\psi^*\psi$) providing the emitter with a choice: to complete the transaction either with the left absorber or with the right absorber so that the beta particle goes either to the right or the left. The probability that either absorber will be chosen is proportional to the strength of the confirmation wave received from it.

The details of what happens to complete the transaction with the selected absorber are distressingly obscure. After having followed out the pseudotime sequence to the reception of the first "echo", we are told that "the emitter responds to the 'echo' and the cycle repeats until the response of the emitter and absorber is sufficient to satisfy all of the quantum boundary conditions ... at which point the transaction is completed" (Cramer 1986, p. 662). Or again, describing the same part of the process: "The exchange then cyclically repeats until the net exchange of energy and other conserved quantities satisfies the quantum boundary conditions of the system, at which point the transaction is complete" (ibid., p. 662). The process of completion achieves some essential aims: all trace of the backwards-moving wave *earlier* than the emission time is eliminated and all of the components of the wave-function representing unactualized possibilities are suppressed.

Cramer is quick to point out that this heuristic account cannot be taken too seriously: "Of course, the pseudotime sequence of the above discussion is only a semantic convenience for describing the onset of the transaction. An observer ... would perceive only the completed transaction, which he could reinterpret as the passage of a single retarded (i.e. positive energy) photon traveling at the speed of light from the emitter to the absorber" (ibid., pp. 662–3). Unfortunately, all of the seeming illumination provided by the account depends on the pseudotime narrative, and any attempt to either eliminate it or to seriously complete it leads to problems.

One problem is that although Cramer presents his account as one of wave collapse, the collapse only takes place in pseudotime. It is not just, as the passage above suggests, that an observer will not *perceive* anything but the completed transaction: there never (in real time) *is* anything but the completed transaction. In real time no uncollapsed wave ever exists, so nothing is ever collapsed.

A second problem is that the most important stage in the transaction only takes place in pseudotime. The "first" echo of the offer wave comes back with an intensity equal to the quantum mechanical probability of the transaction. But in real time there is no "first" or "second" or "third" echo: only the completed transaction, which contains nothing that is proportional to the quantum mechanical probabilities. The beta particle simply either goes to the left absorber or to the right.

A third problem is that nothing in the pseudotime sequence suggests how this could be a *stochastic* theory. All of the mechanisms described, namely the propagation of the wave-function and the "echoing" of the wave-function by potential receivers, are completely deterministic. The atom produces an offer wave which evokes a particular confirmation. The confirmation wave would in turn engender a unique reconfirmation and so on ad infinitum. It is *inconsistent* with this pseudotime sequence that the beta particle go to the left on one run of the experiment and to the right on another, but this is what quantum mechanics predicts and what occurs.

The foregoing three problems are peculiar to Cramer's theory, and might well not arise in some other account which employs backward causation. But there is a final problem or greater scope which will arise for any stochastic theory of this sort. It can be sharply illustrated in Cramer's picture, so let us ignore all of the preceding objections for the moment.

The heart of the transactional interpretation is the idea that an emitter sends a signal forward in time to all potential absorbers and the absorbers send back some sort of confirmation signal. Thus the emitter has, at the time of emission, a sort of catalog of all the possible absorbers. The emitter may also gain some information about the state of the absorbers, information such as the orientation of their polarizers. The emitter then can emit a particular kind of particle, thereby realizing one of the possibilities.

This picture depends crucially on the idea that the absorbers are somehow just sitting out there in the future, waiting to absorb. They represent

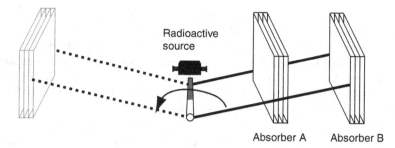

Figure 7.1   Experiment with Movable Absorbers

a fixed field of possibilities among which the emitter can choose. So in our beta particle example, we imagine both the left and the right absorbers signaling back their availability in the future. As we saw, these "echoes" from the future are supposed to determine the probabilities for each possible outcome.

But there is no reason for the absorbers to be fixed in the future, unaffected by everything that happens in the present. The disposition of the absorbers in the future may well depend on how events unfold in the present. So they cannot just passively sit out in the future sending back messages which influence events now.

To make the problem concrete, consider the following embellishment on the beta decay experiment. At t_0 a radioactive source is unshielded, allowing a beta particle to escape either to the right or to the left. Absorber A is situated 1 meter to the right. Absorber B is initially situated 2 meters to the right, but is built on pivots so it can be swung around to the left very quickly on command (figure 7.1). If the particle is absorbed by absorber A it will be so at one second after t_0. If we do not see the particle hit absorber A at that time we will quickly swing around absorber B so it will catch the particle two seconds after t_0 on the left.

It would be easy to set up this experiment so that the quantum mechanical probabilities for the possible results yield a 50 percent chance that the particle will be detected by absorber A and absorber B will remain on the right and a 50 percent chance that the particle will not be detected by absorber A, absorber B will be swung to the left, and the particle will be detected there. At the beginning of the experiment it is not yet determined whether absorber B will end up on the right or on the left. How could Cramer deal with this?

Since there are two possible outcomes, detection at absorber A and detection at absorber B, Cramer's theory seems to demand that there be two confirmation waves sent back from the future, one from absorber A and one from absorber B. These waves should be equal in amplitude to represent the equiprobability of the two results. But the only way that absorber B can send back a confirmation wave, the only way it could receive the offer wave, is if it gets swung over to the left. And if it has been swung over to the left in the future then the outcome of the experiment must already be decided: the particle will not be detected by absorber A. So every time the emitter receives a confirmation wave from absorber B the particle goes to absorber B, even though the magnitude of the echo implies only a 50 percent probability of it being sent there. Cramer's theory collapses.

Any theory in which both backwards and forwards influences conspire to shape events will face this same challenge. If the course of present events depends on the future and the shape of the future is in part determined by the present then there must be some structure which guarantees the existence of a coherent mutual adjustment of all the free variables. A stochastic theory must further guarantee the existence of several alternative consistent solutions and must provide for some means of ascribing probabilities to the possibilities.

Local relativistic theories avoid these problems since solutions to the field equations at a point are constrained only by the values of quantities in one light cone (either past or future) of a point. Thus in a deterministic theory, specifying data along a hyperplane of simultaneity suffices to fix a unique solution at all times, past and future of the plane. Further, the solutions can be generated sequentially: the solution at $t = 0$ can be continued to a solution at $t = 1$ without having had to solve for any value at times beyond $t = 1$. Thus the physical state at one time generates states at all succeeding times in turn.[4]

Local stochastic theories can fare nearly as well. In such a theory, fixing the physical state in the back light cone of a point may not determine the physical state there, but it does determine a unique probability measure over the possible states such that events at space-like separation are statistically independent of one another. Chances operate in such a theory, but only locally. The present moment makes all of its random choices independently and then generates the probabilities for the immediate future, and so on.

Any theory with both backwards and forwards causation cannot have such a structure. Data along a single hypersurface do not suffice to fix the immediate future since that in turn may be affected by its own future. The metaphysical picture of the past generating the future must be abandoned, and along with it the mathematical tractability of local theories. None of this proves that a theory with explicit backwards causation is impossible, but it does caution against undue optimism. The task of building backwards causation into an account of quantum theory remains unaccomplished, despite Cramer's efforts. The best result we have along these lines is that of Wheeler and Feynman for electromagnetism, but that is a deterministic theory. Stochastic examples have yet to be displayed.[5]

Wave collapses along forward light cones come too late and "collapses" along back cones give rise to the problems of ineliminable backward causation. So a wave-collapse theorist might well opt for some intermediate surface. The obvious choice in Minkowski space-time is a flat space-like hyperplane. The pressing question is on *which* flat hyperplane the collapse occurs. Choice of the hyperplane in, for example, an EPR experiment (in which both polarizers are aligned) will determine which observation contains a stochastic element and which is deterministic. According to the chosen hyperplane, one of the measurements will precede the other. The earlier will instigate the collapse, causing a transition of the distant photon from a state of indefinite polarization to a state of definite polarization. The second photon will then be determined either to pass its polarizer or be absorbed. But how is the relevant hyperplane determined?

At this point we may refer back to the discussion of superluminal signals in chapter 4, in which a parallel question about the superluminal signal reception locus occurred. We there proved that no Lorentz invariant method of randomly choosing hyperplanes in Minkowski space-time is possible. So some additional structure must determine along which hyperplane the wave-function collapses. The most obvious options available are these:

1 A preferred family of hyperplanes is built into the structure of space-time.
2 The preferred hyperplane might be determined by the physical state (probably the state of motion) of the source of the particles.
3 The preferred hyperplane might be determined by the physical state (probably the state of motion) of the particles themselves or of the particles and detectors.
4 The preferred hyperplane might be determined by the overall disposition of all matter in the universe.

Each option poses different difficulties and has its own consequences.

If a preferred family of hyperplanes is part of the intrinsic structure of space-time then the fundamental postulate of the Theory of Relativity is false. Relativity postulates that the metrical structure is the *only* intrinsic spatio-temporal structure and that, more specifically, nothing intrinsic to space-time picks out a preferred reference frame. A family of hyperplanes distinguishes one frame over all others, namely the frame in which the hyperplanes are simultaneity slices. This would not demand the elimination of any relativistic structure, but would undercut the relativistic democracy of frames. In a relativistic milieu, absolute simultaneity implies a unique privileged frame, allowing the definition of Absolute Motion and Rest.[6] So option 1 can be adopted only by those who are willing to throw out the relativistic world-view.

Option 2 faces several difficulties. First, it is a bit bizarre, in that the source of the particles may have long been destroyed before any measurement event. So even though the source may pick out a preferred frame at the moment of emission (for example, its rest frame), it is hard to see how information about that frame could be carried forward to the moment of collapse. A more serious problem is that the Bell inequalities may be violated by sets of particles that don't arise from a single source. Bernard Yurke and David Stoler (1992a, b) have argued that triples of photons that arise independently of one another will display the behavior described by Greenberger, Horne, and Zeilinger (see appendix A). If this is so, appeal to the state of the source will be bootless.

Option 3 avoids the problems of option 2 but faces an even more devastating criticism. This option looks to the state of motion of the particles themselves or of the detectors to define a preferred hyperplane for the collapse. But trajectories of pairs of particles or detectors in Minkowski space-time may simply not pick out any preferred frame.

This result is surprising since any set of particles in a *classical* space-time will determine a preferred frame of reference, namely the center-of-mass frame. At any moment, a collection of classical particles has a unique center of mass, and the trajectory of that center of mass through time defines a frame of reference. Further, if the particles are subject to no outside forces, the center of mass travels inertially and the center-of-mass frame is an inertial frame. If one has been brought up on classical physics, then, one expects any collection of particles to pick out a frame of reference in this way.

But the definition of the center of mass of a system requires fixing the positions of the particles *at a moment*. This is unproblematic in the classical regime, but is precisely the bone of contention in Relativity. Different choices of simultaneity yield different calculations of the position, and trajectory, of the center of mass. Indeed, in some relativistic situations *every* inertial frame would regard itself as a center-of-mass frame. This is most easily seen by an example in a two dimensional space-time. Consider the situation depicted in figure 7.2. The space-time contains two objects (they could be particles or detectors) which travel along time-like hyperbolic trajectories.[7] It is evident by the bilateral symmetry of the diagram that the unprimed frame will regard itself as a center-of-mass frame: the center of mass of the two particles will, according to its definition of simultaneity, always lie on the central time axis. But if we switch to the primed framed, using its definition of simultaneity, the center of mass of the particles is moving to the right (bisect the dotted lines to find the center of mass). In fact, the primed frame is *also* a frame in which the center of mass of the system is at rest. This is easily seen since the hyperbola is a Lorentz invariant locus. A Lorentz transformation into the primed coordinates will, therefore, yield a diagram that looks just like figure 7.2.

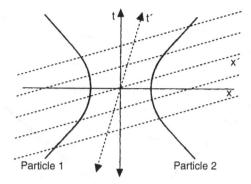

Figure 7.2 Center-of-mass Frames

Furthermore, as David Albert has pointed out to me, not only may there be relativistic worlds with *many* center-of-mass rest frames, there can be worlds with *no* such frame. Most obviously, a world which contains but one photon will have no frame in which the center of mass is at rest. Unlike the situation in classical physics, then, distribution of matter in a relativistic space-time may not serve to pick out any unique frame of reference. Options 3 and 4 therefore cannot always solve the problem.[8]

In sum, relativistic theories of wave collapse do not yet exist. The search for such theories leads either into the acceptance of ineliminable backwards causation, as in Cramer's theory, or in the postulation of a preferred reference frame in space-time. The former possibility presents formidable technical difficulties, the latter is a *de facto* rejection of Relativity. Lorentz invariance is not to be had cheaply or easily if one follows this route.

At the time this third edition was in preparation, the situation described above was changed by the development of a completely Lorentz invariant collapse theory, relativistic flashy GRW, described in chapter 10.

Lorentz Invariance: Hyperplane Dependence

Since wave collapse in non-relativistic quantum mechanics is instantaneous, the most natural suggestion for modeling it in a relativistic theory is by the use of flat space-like hyperplanes, as these represent moments in time according to the various inertial frames of reference. We have just seen that the supposition that the wave-function collapses along some *unique* hyperplane is difficult to reconcile with Relativity since there are no obvious Lorentz invariant means of selecting a preferred hyperplane. This difficulty is a purely theoretical one, in that the choice of one hyperplane or another will not affect the statistical predictions which the theory makes. No observations could help

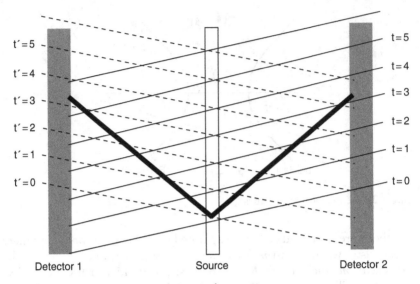

Figure 7.3   ERP Experiment and Two Reference Frames

us determine which hyperplane contains the collapse, but equally no observation could refute the claim that a particular hyperplane was used. So one might be moved by the spirit of relativistic democracy to allow every inertial frame to employ its own simultaneity slices as the preferred hyperplanes. In every reference frame, collapses are instantaneous *in that reference frame*, and each frame thereby is able to tell a physical story which makes all of the right predictions.

Of course, the stories the various observers tell may differ radically from one another. Consider an EPR situation in which the two distant polarizers are set in the same direction. The detection events are at space-like separation, so different frames will disagree about their time order. According to the unprimed frame in figure 7.3, the particle on the right is detected before the one on the left, while according to the primed frame the left detection comes first. The unprimed frame recounts events as follows: a pair of photons are created at t = 0 in the singlet state. Neither photon has a definite polarization. At t = 3.2 the right-hand photon reaches detector 2. At that point a completely indeterministic event takes place. In this case, the photon passes the detector, causing the wave-function for the pair to collapse. After t = 3.2 the left-hand photon is in a state of definite polarization, oriented in the direction of the polarizer on detector 2. When the second photon reaches its detector at t = 5.5, a completely deterministic event takes place. Being already polarized in the correct direction, the left-hand photon will certainly pass its detector.

The unprimed frame gives exactly the same account, save that "t" is replaced by "t'" and "right" is interchanged with "left." According to that frame, the stochastic event takes place at detector 1 and the deterministic one at detector 2.

Can we allow these two apparently contradictory versions of events to peacefully co-exist? At first glance this would seem to be impossible. The two stories disagree not only on the spatial and temporal structure of the events, as we would expect, but also on seemingly non-spatio-temporal attributes such as the polarization states of the photons. How could it be correct both to say that the photon on the right has no polarization before measured and that it does have one? Or that the firing of detector 1 is both a deterministic and a stochastic process? Are there no limits on how the versions generated by the various reference frames may disagree?

We already know that the relativistic world looks very different according to one's reference frame. But Relativity also reveals some of the apparent contradictions between frames to be merely matters of equivocation. The unprimed frame says that the right-hand photon is detected before the left while the primed frame has it the other way around. How could they both be right? In this case the answer is clear: they are simply talking about different things. The unprimed frame notes precedence in its t-coordinate, which we might call "time," while the primed frame is concerned with precedence in its own t-coordinate, which we could call "primetime." There is no more contradiction between saying that the right detection event precedes the left in time but follows it in primetime than there is in saying that Idaho precedes New Jersey in geographical area but follows it in terms of population. Many apparent contradictions in Relativity are immediately resolved once one disambiguates terms this way.

Other problems are somewhat more tenacious. In chapter 2 we discussed the problem of the relativistic car going through the tunnel. In the car's own reference frame it is longer than the tunnel while according to the tunnel the car is shorter. How could two reference frames disagree on such a primitive property as containment? Surely they both must be in accord on whether or not the ends of the car are inside or outside of the tunnel!

Revisiting our space-time diagram of the situation (figure 7.4) we can see that the solution lies in more than just the application of the Lorentz transformations. The paradox is resolved by noting that when shifting between frames we have not just re-coordinatized points, we have also changed the hyperplane under consideration (this was also part of the moral of the shrinking bar calculation). If we don't shift hyperplanes then the car and tunnel frames will make notably similar pronouncements.

Thus, for example, both the car and the tunnel reference frame will agree that *along the* t = *constant hyperplane* the front and back ends of the car are both inside the tunnel, and that *along the* t' = *constant hyperplane* they are both out. (The frames will still disagree about some matters, for example *how far* it is from the front of the car to the edge of the tunnel, but this disagreement is

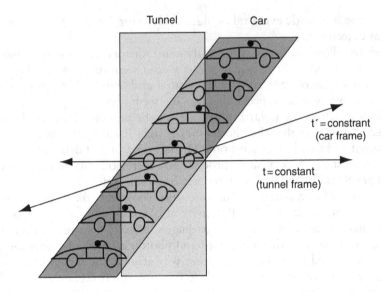

Figure 7.4   The Car and Tunnel Again

of the time/primetime variety.) So one should not confuse two very different changes which may occur when moving from one reference frame to another. On the one hand, the spatial and temporal distances between points will change in accord with the Lorentz transformation. But also, one often switches the relevant hyperplane under consideration from a simultaneity slice in the old frame to one in the new frame. These changes are quite independent of one another: one can apply the Lorentz transformations to a single locus of points along a given hyperplane or one can change hyperplanes in a given reference frame, avoiding the Lorentz transformation entirely.

This dual aspect of the transformation between inertial frames in Special Relativity produces a serious disanalogy with the Galilean transformation. We saw that a theory is Galilean invariant if a Galilean transformation at $t = 0$ followed by time evolution according to the dynamical laws always gives the same result as the temporal evolution followed by the Galilean transformation. But the analogous relativistic transformation at $t = 0$ is commonly taken to change not only the state of the system but also the hyperplane along which that state is evaluated. The Lorentz transformations alone cannot generate this new state since the original state is defined along a different locus of points as the transformed state.

Once one pulls these two sorts of changes apart, it becomes obvious that it is not the Lorentz transformation which is generating the contradictions between the two stories in the EPR example, but the change in hyperplanes. The $t = 5$ hyperplane, for example, is one which comes *after* the measurement on detector 2 and *before* the measurement on detector 1. All reference frames would agree to this much. Therefore if one were to ask for the state of the photons *relative to the* t = 5 *hyperplane, all* reference frames would give the same answer: the right-hand photon has already been measured and the left photon is in a state of definite polarization. Similarly, all reference frames agree that *relative to the* t' = 5 *hyperplane* the right-hand photon is in a definite state of polarization, induced by wave collapse from the measurement on the left.

This idea for constructing a Lorentz invariant theory of wave collapse has been developed in great detail by Gordon Fleming in his *hyperplane dependent quantum field theory* (Fleming 1985, 1986; Fleming and Bennett 1989; see also Giovannini 1983, Aharonov and Albert 1984, Dieks 1985). Much of Fleming's work is concerned with the technical details of implementing the Lorentz transformation on a hyperplane, but for our purposes we need only consider the fundamental ontological picture. The crux of the position is announced in the name of the theory: hyperplane dependence. Fleming achieves Lorentz invariance by insisting that the quantum state of a particle depends crucially on the hyperplane one considers. Even properties such as polarization, which don't seem to have any spatio-temporal dependence, are defined only relative to a hyperplane. So if one asks what the state of the right-hand photon is immediately prior to the measurement at detector 2, Fleming would regard this question as incomplete. One must further specify the hyperplane one has in mind. If one chooses the t = 3 hyperplane then the photon is in a state of no definite polarization. If one chooses the t' = 5 hyperplane *which intersects the photon's worldline at the very same point*[9] then the photon is in a definite state of polarization. Even though the photon may be localized near a point, its polarization is not determined by just indicating that point: one must choose a hyperplane through the point as well.

Insisting on the hyperplane dependence of various properties can make it relatively simple to produce Lorentz invariant theories. In the case of polarization, the hyperplane dependent states described above are directly Lorentz invariant. The primed and unprimed frames assign the very same state to the photon on the right *evaluated along the* t = 3 *hyperplane*. They both also assign the same state to the photon *evaluated along the* t' = 5 *hyperplane*. The seeming contradiction between descriptions in different reference frames dissolves.

Of course, one might worry that the original contradiction just reappears within each reference frame taken alone. Does it really make sense to say that a photon at a particular location has one polarization state when thought of as lying along one hyperplane and another when thought of as lying along a different one? Or that a detection event is stochastic when one assigns it to one hyperplane but deterministic when assigned to another? The original difficulty in picking a hyperplane for wave collapse has been swamped by a radical generosity: in Fleming's theory there are wave collapses along *an infinity* of hyperplanes that pass through *either* detection event. Has Nature really been so profligate with collapse events?

We are here faced with a radically new ontological conception of the world, so it is hard to find the appropriate bearings from which to make an evaluation. The theory is consistent and Lorentz invariant, but not surprisingly it shocks our commonsense intuitions. It also shocks intuitions rooted in the study of classical physics. Most interestingly, it even shocks intuitions which are formed by acquaintance both with Relativity and with non-relativistic quantum mechanics. So the best we can do here is to delineate clearly and carefully just how the hyperplane dependent theory employs a new picture of physical reality.

The first step is to eliminate all reference, implicit or explicit, to human intentional action. It is easy to fall into locutions such as those used above: the photon *considered as lying along such-and-such a hyperplane* has a definite polarization, or the photon *evaluated along such-and-such a hyperplane* is unpolarized. We don't want to give the impression that physical states of photons depend on any human acts of either *consideration* or *evaluation*. The photon which lies at the intersection of the $t = 3$ and $t' = 5$ hypersurfaces lies on both hypersurfaces, irrespective of any human cognitive acts. As such, it has both polarization states, one relative to one hypersurface, the other relative to the other.

The language of relations, of the photon having a property relative to a hypersurface, makes the situation sound a bit more prosaic. Socrates can be tall relative to Theaetetus and short relative to Plato without thereby engendering any paradoxes or puzzlement. But hyperplane dependence is more unfamiliar, and stranger, than this analogy suggests. First, it is obvious that tallness and shortness are relations, not properties which an object could have in and of itself alone. If the universe contained but one object, that object could not in principle be either short or tall. But polarization is not such an intrinsically relational matter. In a universe with only one photon, the photon could have a definite state of polarization. Indeed, one has the strong intuition that whether or not a photon is polarized should be a matter purely of the intrinsic state of it, independent of any considerations about hyperplanes.

This intuition about the intrinsic nature of properties was well described by Einstein in the passage which was cited in the chapter 1:

> An essential aspect of this arrangement of things in physics is that they lay claim, at a certain time, to an existence independent of one another, provided these objects "are situated in different parts of space." Unless one makes this kind of assumption about the independence of the existence (the "being-thus") of objects which are far apart from one another in space – which stems in the first place from everyday thinking – physical thinking in the familiar sense would not be possible. (Born 1971, pp. 170–1)

Einstein's intuition was that properties such as the polarization of the photon should be a matter purely of the physical state of things in a small region containing the photon and should be independent of anything going on at space-like separation. It should certainly be independent of whether one happens to choose a hyperplane which lies before or after some distant measurement event.

This aspect of Einstein's world-view has been remarked and commented on by several philosophers. Don Howard (1985, 1989), as we noted in chapter 4, has argued extensively that Einstein subscribed to two quite distinct theses about the world: separability and locality. Paul Teller (1989) has argued that the ultimate moral of Bell's work should be rejection of the idea that all relations simply supervene on the non-relational properties of the relata (as Socrates's tallness relative to Plato is determined entirely by the non-relational height of each). Howard's and Teller's ideas can be both illustrated and clarified by Fleming's notion of hyperplane dependence.

Einstein's world-view held that each region of space-time has its own intrinsic physical state, and that the entire and complete physical state of the universe is specified once one has determined the intrinsic state of each small region. This ontological doctrine may be called "separability." Einstein also believed that events at space-like separation from a region could have no influence on the physical state assigned to that region. Let us call that doctrine "locality."

We have already seen (chapter 4) that Howard mistakenly identifies locality and separability with Jarrett's locality and completeness, i.e. with parameter independence and outcome independence respectively. (This identification is mistaken since one can, for example, construct completely separable theories which violate outcome independence.) The burden of the last two chapters was to show that any theory which predicts violations of Bell's inequality (at space-like separation) must violate locality as defined above. The theory of hyperplane dependence displays violations of both locality and separability.

The violations of locality are evident since Fleming's theory contains wave collapses along space-like hyperplanes. Due to these collapses, the physical state of one particle may be determined by the stochastic outcome of a measurement at space-like separation. More interesting is the violation of separability. As we have seen, the polarization state of a photon in this theory is not an intrinsic property simply of the space-time region the photon occupies.[10] The photon has a polarization state only as a component of a larger whole, a complete hyperplane.

Hyperplane dependence should also be distinguished from another form of non-separability. The singlet state which the photons start out in is non-separable in the technical sense of not being a product of polarization states of the individual particles. That is, the polarization of the pair is something over and above the sum of the states of the parts. This is part of the inspiration for Teller's claim. But in Fleming's theory, whether or not the polarization of a photon is, in this technical sense, separable is itself a hyperplane dependent matter. For example, the photon on the right in figure 7.3 is in a (technically) separable state if one evaluates along the $t' = 5$ hyperplane but in a non-separable state along the $t = 3$ hyperplane.

Teller's claim that classical intuitions presuppose that *all* relations supervene on the intrinsic properties of the relata (that they are, in philosophical vocabulary, all internal relations) is surely too strong. No one ever thought that all relations are internal since it is obvious that spatial and temporal relations are not.[11] No amount of description of the properties of two objects will reveal how far apart they are. But one could believe, as Einstein apparently did, that spatio-temporal relations are the *only* fundamental external relations. The thesis that the entire physical state of the universe must be specified once one gives the intrinsic physical state of each region of space-time follows from this supposition.

The theory of hyperplane dependence contradicts Einstein's supposition in a particularly radical fashion. It is not just that there are some relations which don't supervene on the intrinsic state of the relata but that the relata *don't have any intrinsic state at all*. It is not just that the whole is more than the sum of the parts but that the parts can't even be defined apart from the whole.

Much of modern science has been predicated on the postulate that compound systems are to be understood by analyzing them in terms of their parts. Teller has correctly identified one of the disturbing features of quantum theory, the feature David Bohm has called "undivided wholeness." But only Fleming's formalism makes clear the consequences of this wholeness given Relativity. A single photon is part of *many* wholes, as many as there are space-like hyperplanes through it. The photon may be attributed different characteristics depending on the hyperplane chosen, and in a Lorentz invariant theory there are no grounds for privileging one choice over another. The consequent rejection of Einstein's notion of separability is therefore deeper and more shocking than merely the acceptance of some new external relations. And merely counseling the recognition of such relations, as Teller does, does not address the underlying question of consistency with Relativity.

The theory of hyperplane dependence achieves the goal of a Lorentz invariant theory. It does so via a fundamental rejection of the ontological foundations of both common sense and of classical physics. Indeed, the rejection is more radical than either Relativity or non-relativistic quantum theory suggest. Photons in non-relativistic quantum theory may be in strange states of indefinite polarization, but at least they are perfectly determinate strange states. Photons in Fleming's theory can be both polarized and not polarized.

Measurement events can be both deterministic and stochastic. It all depends on the particular hyperplane to which one assigns the photon. And Nature assigns the photons even-handedly to all the hyperplanes at once.

Lorentz Invariance: Non-Collapse Theories

Given the great difficulties involved in constructing a Lorentz invariant theory of wave collapse, one might suspect that a quantum theory without wave collapse would be more easily amenable to relativistic generalization. Further, we already have at hand a non-collapse theory, namely Bohm's theory. So perhaps abandoning orthodox quantum theory with wave collapse for Bohm's theory will alleviate our problems.

Unfortunately, our hopes are quickly dashed. Even without any collapses, non-relativistic Bohmian mechanics makes cardinal use of simultaneity in a way that resists any relativistic formulation. A brief review of the structure of the non-relativistic theory will illuminate the problem.

Recall that Bohm's theory is a deterministic theory: no irreducibly random events happen in this ontology. But we know that the behavior of each photon cannot be determined merely by the initial state of the pair and the setting of its own polarizer. If it could, then a completely local theory could be constructed and Bell's inequality could not be violated. So at least one photon's behavior must be determined not only by the initial state and the setting of its own polarizer, but also by the setting of the distant polarizer. That is, for at least some initial states and some polarizer settings, whether or not the right-hand photon passes its analyzer must depend on how the left-hand analyzer is set.

But *when* is the left-hand setting relevant? If the left-hand photon is put into storage and not examined for ten years, the observation made in the future can have no influence on the right-hand observation made today (Bohm's theory has no backward causation). In the non-relativistic theory the answer is straightforward: the measurement on the left can influence the outcome on the right only if it is made *before* the measurement on the right. If the right-hand observation comes first then the result will depend only on the initial state, although in this case the right-hand setting may help determine the left-hand result. Bell (1987, pp. 131–2) provides a complete technical presentation of this situation. He also shows that this dependence on the distant measurement obtains only when the particles are in an entangled state.

We can now see why the transition to a Lorentz invariant version of the theory cannot be smooth. In the non-relativistic theory the behavior of one of the pair will be affected by the distant setting only if its twin is observed *first*. But in Minkowski space-time one cannot univocally denominate either

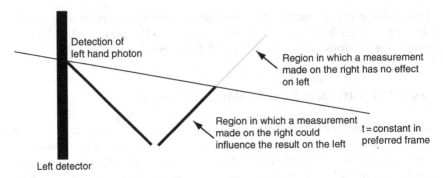

Detection of
left hand photon

Region in which a measurement
made on the right has no effect
on left

Region in which a measurement
made on the right could
influence the result on the left

t = constant in
preferred frame

Left detector

Figure 7.5 How Distant Dependence Privileges a Frame

of two space-like separated measurements the first. Which measurement comes first depends on the reference frame chosen. So if all the reference frames simply adopt the non-relativistic theory, each using its own version of simultaneity, they will tell stories that are inconsistent with one another. According to one frame, for example, the result on the right would have been different had the left-hand detector been set differently (since in that frame the left-hand observation came first) while according to another frame, in which the right-hand measurement was made first, changing the left-hand detector would have made no difference.[12]

Given that the contradiction considered above involves counterfactuals, it is interesting to speculate whether one could construct a theory in which all reference frames agree on what *actually* happens but disagree on what *would have* happened under various circumstances. Would such a theory be Lorentz invariant or not? One might also consider trying to construct a hyperplane dependent version of Bohm's theory, but, as David Albert has noted (p.c.), the theory as it is resists this.

The underlying problem for a relativistic theory, then, is not wave collapse *per se* but rather the non-local dependence of one measurement result on the distant setting. In collapse theories that dependence is secured through the collapse; in Bohm's theory it is mediated through the uncollapsed wave-function. Wherever there is such a dependence then (in the absence of backwards causation *à la* Cramer) there must also be a cut-off point beyond which, e.g., the setting on the right can no longer have an effect on the left, for the distant measurement can be postponed indefinitely, or never performed at all. But that cut-off point defines a preferred frame of reference, namely the frame in which the cut-off is simultaneous with the measurement on the left (figure 7.5). And we are back to the problem of picking out that preferred frame in a Lorentz invariant way.

One still might be puzzled at just how a preferred notion of simultaneity gets smuggled into Bohm's theory in spite of the absence of the manifestly instantaneous process of non-relativistic wave collapse. Part of the answer depends on the recognition that the wave-function itself is not defined over, and therefore does not exist in, space-time.

The non-relativistic quantum mechanical wave-function is defined over the *configuration space* of a system. A single point in configuration space represents an entire configuration for the particles in the system, that is, it specifies where all of the particles are. So, for example, consider a very simple system consisting in two distinguishable particles, A and B, and two boxes 1 and 2. Each of the particles could be in either box. The configuration space of the system would have four regions: one representing both particles in box 1, one representing particle A in box 1 and particle B in box 2, one representing particle A in box 2 and particle B in box 1, and one representing both particles in box 2. The wave-function for the system is not defined over the physical space composed of box 1 and box 2 but over the configuration which has all four of these regions.

This use of configuration space is essential to the prediction of the sorts of correlations we have been concerned with. For example, a wave-function for the system just described might have half of its amplitude in the region representing both particles in box 1 and half of its amplitude in the region representing both particles in box 2, with zero amplitude elsewhere. Under the orthodox interpretation of quantum theory, this would represent a situation in which neither particle has a definite location: upon measurement, either could be found in either box. But although neither particle has a definite location, their locations are nonetheless correlated! That is, there is no chance of finding them in different boxes: observing either particle to be in box 1 guarantees that the other will be found there too.

One cannot capture this correlational information if one thinks of the wave-function as inhabiting ordinary space, for the best one could say of such a wave-function is that half of its amplitude for each particle is in box 1 and the other half is in box 2 (i.e. there is a 50 percent chance for each particle to be found in either box). But this information is compatible with a very different correlational structure. The wave-function could, for example, have zero amplitude in the regions of configuration space which correspond to the particles being in the same box, with a 50 percent chance that particle A is in box 1 and B in box 2 and a 50 percent chance that A is in box 2 and B in box 1. There is then still a 50 percent chance for each particle to be in either box, but no chance that they are in the same box.

The fact that the wave-function exists in configuration space rather than physical space is often overlooked because for a single particle configuration space is *isomorphic* to physical space (one specifies the complete "configuration" of a one-particle system by saying where it is in space). One-particle problems, such as the infamous two-slit experiment, can therefore be analyzed as if the wave-function were a classical field in space. But as soon as more than one particle is involved this analogy becomes untenable. In general, a system consisting in n particles inhabiting an m-dimensional space will have a wave-function defined over an (n × m)-dimensional configuration space.

One of the problems which faces relativistic quantum mechanics is that the notion of a configuration, as used in the non-relativistic version, already presupposes simultaneity. Configurations are *configurations at a time,* they specify where all of the particles in a system are *at a given moment.* So the very notion of a configuration is not a Lorentz invariant concept.

As an obvious example, in the car-and-tunnel problem of figure 7.4, in the car's frame of reference there is a configuration of the car and tunnel in which both ends of the car are sticking out of the tunnel. In the tunnel's frame of reference no such configuration ever exists. So given the way simultaneity is presupposed in the very notion of a configuration, together with the fact that the wave-function is defined over configuration space, it is not surprising that some notion of simultaneity would find its way inextricably into Bohm's theory. Any theory which takes the wave-function seriously may well face similar problems, whether the wave-function collapses or not.

The problem of Lorentz invariance should now seem quite formidable. There is, however, one non-collapse theory which runs into no difficulties at all with Lorentz invariance. Indeed, the theory is completely consistent with any conceivable relativistic constraint because it is *a completely local theory.* There is no sort of non-local, superluminal action or influence at all.

The preceding revelation may well shock, even outrage, the patient reader. The course of this book thus far has been one sustained argument that the predictions of quantum mechanics cannot be reconciled with any local theory. What then of all of these arguments?

The theory we are about to consider does not show that our argumentation has been invalid, but rather highlights a certain supposition which has underlain all of our investigations. It is a supposition so fundamental that it is hard initially to imagine how one could abandon it. From the very beginning we have simply taken for granted that *measurements made on particles have determinate results.* In our illustrative game, this amounts to demanding that every question asked be given a single univocal answer; physically it means that every photon either passes its polarizer or is absorbed by it. Of all of our strange speculations, we are about to embark on the strangest: that physical measurements may not have such determinate results at all.

The route to this bizarre hypothesis is surprisingly straightforward. We have already seen that the collapse of the wave-function, although central to the orthodox interpretation of quantum mechanics, is grafted onto the formalism of the theory in a completely *ad hoc* way. The orthodox theory has no precise characterization of how or when wave-collapse occurs. Collapse is supposed to be associated with "measurements," but no exact characterization of a measurement is forthcoming. This interpretive problem is most easily avoided by denying the collapses altogether, hence the appeal of Bohm's theory.

But the denial of collapses carries its own interpretive burden. The fundamental problem has been long recognized and has been commonly illustrated by Schrödinger's example of the cat. Adapting the point to our experimental situation, the puzzle is as follows. We know that without any wave collapses, the incoming photons reach their detectors in a state of indefinite polarization. They are neither definitely disposed to pass the polarizer and set off the detector nor to be absorbed by the polarizer. If one simply allows the wave-function to evolve in accord with Schrödinger's equation, this indeterminateness infiltrates into the measuring device itself. Just as the wave-function before the observation represents a photon which is in no definite state of polarization, so the wave-function after the interaction represents the detector as being in no definite state of detection. That is, the wave-function of the detector evolves into a superposition of two states, one in which the photon is absorbed and another in which it is passed. According to the wave-function itself, the photon has been neither definitely absorbed nor definitely passed.

The fundamental interpretive problem of quantum theory is accounting for this consequence. The orthodox interpretation does so by denying that the detector ever gets into such an odd state. Somewhere Schrödinger evolution must break down with the collapse of the wave-function. Bohm's theory avoids the problem by denying that the wave-function is all there is in the world. Although the wave-function may evolve into this odd state, the detector itself is composed of particles which are always in a perfectly determinate configuration. After the measurement this may be a configuration of a detector which has been triggered or of one which has not. But if one denies that the wave-function collapses and also denies that there is more to the physical world than is represented in the wave-function both of these avenues are closed off. The detector can neither fire nor refrain from firing.

The problem with such a theory, of course, is reconciling it with our experience. Anyone who watches the detector will either see it fire or not. Our experience does not come in strange indeterminate states. How then can such determinate psychical experiences fit into a world of indeterminate physical states?

Attacking this problem radically at precisely this point, David Albert and Barry Loewer have developed the *Many-Minds Interpretation* of quantum mechanics (Albert and Loewer 1988, 1989; Albert 1992). Albert and Loewer postulate that detectors do get into the funny indeterminate states described above. And when people look at the detectors their bodies get into similar funny indeterminate states, states in which, for example, their retinas have neither definitely received photons from the detector nor definitely not received them. Their brains also get into strange indeterminate states. But their minds do not.

More precisely, Albert and Loewer postulate that minds are never in superpositions of different belief states. So if your body manages to get into a superposition of seeing the detector fire and not seeing it fire, your mind will nonetheless either experience it as firing or as not firing, sometimes the one and sometimes the other. There is a stochastic connection between the body and the mind. The brain can, as it were, offer the mind a set of possible experiences and the mind chooses one of these.

Albert and Loewer's theory is actually even more radical than this sketch indicates: they associate with every sentient body not one mind but an infinite number, minds which may be in different experiential states. But for our purposes this additional fillip is inessential, so let's not burden ourselves with it now.

The fundamental claim of the many-minds theory is that definite outcomes *only take place in individual minds*. Furthermore, the sorts of correlations with which we have been concerned *exist only in individual minds*. Since there are no relevant correlations between space-like separated events, such as the responses of measuring devices, there is no problem of non-locality.

Let's work through our standard experimental set-up in some detail. For simplicity, suppose that the polarizers are set in the same direction so we expect perfectly correlated results. First, suppose the experiment is run with no sentient beings around. Each detector goes into an indefinite state: neither photon is either absorbed or passed. The overall quantum state does code up the relevant correlations in that none of the wave-function corresponds to a state in which the detectors give different results. But as with the particles in boxes described above, the correlation in the wave-function exists without there being definite results which are correlated.

Now suppose an observer appears on the scene and examines one of the detectors. The observer's body goes into an indefinite state: his eyes record neither one definite result nor the other. But the observer's mind does have the impression that some definite result occurred. The observer "sees" the photon as either having passed or having been absorbed.

At this point in the story absolutely nothing happens at the distant detector. It remains as it was, in an indefinite state. The fact that our observer has "seen" one result does not collapse the wave-function, or change the distant physical situation in any way.

Now suppose our observer wanders over to the second detector. No definite result occurred there, so the observer's brain does not register either particular result. But because of the correlation in the wave-function, the wave-function of the observer's brain has no component which corresponds to seeing the detectors yield different results. So *the observer's mind will perceive the detectors as having given agreeing results*. But this correlation between perceptions only occurs when the observer has interacted with both detectors. Since each of these interactions is a purely *local* affair, no superluminal effects need be invoked.

The situation becomes rather more complicated if more sentient observers are introduced, but one must always remember that the only *relevant* correlations, the only *observed* correlations, are correlations between perceptions in a *single* mind. So for example, suppose that there is an observer stationed at each detector. Each observer will perceive a particular result. These results might not agree. The right-hand observer might see his photon pass his polarizer while the left-hand observer sees his be absorbed. This seems to contradict the quantum mechanical prediction of perfect correlation, but recall that so far no correlation has been observed by anyone. Neither of our experimentalists would yet see anything untoward. To get an observed correlation they must communicate with one another.

The key here lies in the fact that the minds of our two observers cannot directly communicate with each other. One experimentalist can only interact with the *body* of the other, and the bodies continue to be in indefinite states. Indeed, the body of each experimentalist is in no different state than the detector he looked at. So when the observer who saw the photon pass asks his colleague to report, he will perceive the answer to be "I saw my photon pass." The observer who saw the photon be absorbed will also have the impression that both detectors reacted similarly, but he will have the impression that both photons were absorbed. In each individual mind the quantum predictions are upheld, even though they have quite different views about exactly what happened.

Similarly, in a long run of experiments designed to test for violations of Bell's inequality, every observer is overwhelmingly likely to see the inequality violated, but any pair of observers are also likely to perceive a different sequence of outcomes.

The reader will doubtless have noticed that when two observers talk to one another, they are quite likely to come away with the impression that their interlocutor reported observing a course of events that the interlocutor's mind never experienced. That is, many conversations will take place with "empty hulks": bodies seemingly behaving in intelligent ways even though there is no corresponding mind behind the behavior. It is in part to avoid such hulks that Albert and Loewer postulate an infinity of minds to each body. For every sequence of events an observer could see, some mind does see it.

Choose Your Poison

We have come to the end of our discussion of Lorentz invariance. Embedding quantum theory into the Minkowski space-time is not an impossible task, but all the available options demand some rather severe sacrifices. Bohm's theory and orthodox collapse theories seem to require the postulation of some preferred set of hyperplanes, ruining Lorentz invariance and jettisoning the fundamental ontological claims of the Special Theory. Lorentz invariant theories might be developed using explicit backwards causation, so that the birth of every pair of correlated particles is already affected by the sorts of measurements to which they will be subject, perhaps hundreds of years later. Cramer has attempted something in this direction, but his own interpretation fails. One can achieve Lorentz invariance in a theory with instantaneous collapses by embracing a hyperplane dependent formalism, but the exact ontological implications of such a view are rather hard to grasp. (The view also inherits from orthodox quantum theory the problem of explaining which interactions induce collapses.) Or finally, one can both avoid collapses and retain locality by embracing the many-minds ontology, exacting a rather high price from common sense.

None of these choices is particularly appealing. This is not to disparage in any way the efforts of those who have developed these theories. They have seen a difficult and fundamental problem and they have taken it head on. Perhaps other, equally surprising options will come to light. But the common thread that runs through all of these proposals is that no results are to be had at a low price. Indeed, the cost exacted by those theories which retain Lorentz invariance is so high that one might rationally prefer to reject Relativity as the ultimate account of space-time structure.

Notes

1   If one formulates Bohm's theory in terms of the quantum potential then it will look more like a Newtonian theory, and Galilean invariance will be shown differently.

2   There is a rather artificial way to collapse the second case into the first. Thus, Bohm's equation of motion for the position q of a single particle is

$$\frac{dq}{dt} = hIm\frac{grad\psi}{\psi}(q).$$

Both $\frac{dq}{dt}$ and ψ are not invariant under a Galilean transformation. But one can rewrite the equation as

$$\left(\frac{dq}{dt} - hIm\frac{grad\psi}{\psi}(q)\right) = 0,$$

showing that the quantity

$$\frac{dq}{dt} - hIm\frac{grad\psi}{\psi}(q)$$

is invariant under the transformation. Obviously, any invariant equation couched in terms of non-invariant quantities can similarly be transformed into an equation which states that some invariant quantity is identically zero. But it is hard to see how this formal trick provides any real insight, since one does not want to interpret every state in which the invariant quantity equals zero as a vacuum state.

3 In the case of a massive particle such as a beta, the "collapse" is not along the back light cone but along a backwards time-like curve, essentially the trajectory of the particle. This avoids some of the objections brought by Yakir Aharonov and David Albert against using light cone collapses for massive particles (Aharonov and Albert 1981, pp. 24ff.). As mentioned above, Cramer's theory is best seen as one without any collapses since the uncollapsed states never exist. It should not therefore be confused with the theory of Hellwig and Crause (1970) which postulates wave collapse along back light-cones for *all* particles. It is this theory that Aharonov and Albert directly criticize.

4 Questions of determinism are not quite so simple as they have been presented here, although local relativistic equations in Minkowski space-time are very nearly ideal deterministic theories. See Earman 1986 for all the relevant caveats and qualifications.

5 It is also notable that in the electromagnetic case the relevant fields are defined on, and propagate over, space-time. The wave-function is defined on configuration space. Cramer does not seem to take account of this, writing always as if his offer and confirmation waves were simply being sent through space. Any theory which seeks to make the wave-function directly a medium of backwards causation ought to take this into account.

6 Ironically, it is the relativistic metrical structure itself that allows a privileged notion of simultaneity to define a notion of absolute motion. Galilean space-time has a single well-defined simultaneity relation but no absolute motion. So Relativity plus absolute simultaneity yields Newtonian Absolute Space, but not via the Newtonian postulate of individual parts of space enduring through time. It is rather shocking to see a structure which supports absolute motion but does not define it as change of absolute place.

7 If one considers the objects in figure 7.2 to be the particles, then the experiment evidently is not using photons. But violations of the Bell inequality are also produced by electrons and other massive particles, using spin instead of polarization.

8 When a unique center-of-mass frame does exist, though, one can construct a Lorentz invariant theory. See Smith and Weingard 1987 for such a construction.

9 There is some degree of idealization involved in this description. The photon cannot have an entirely precise position in free flight according to the usual interpretation of the wave-function, so there may not be any exact worldline of the photon. In some circumstances, this fact looms large. If, for example, we were interested in measuring the exact position of the photon, we can locate it exactly on at most one hyperplane. That is, a particle with an exact position on one hyperplane will not have an exact position on any other, *even on hyperplanes which intersect the observed "position" of the particle.* There is no Lorentz invariant notion of exact position in relativistic quantum theory. Fortunately, we have chosen to measure a property of the photon, the polarization, which is not

coupled to the position. So we have no real fundamental problems, although it might be best to think of the photon as localized to a region rather than a point.

10   In referring to the "polarization state" of the photon I do not mean to presuppose that the photon *by itself* necessarily has any state at all. There is some dispute whether a member of a pair in an entangled state should itself be ascribed a so-called reduced state or should only be ascribed a joint state with the second photon. I incline to the latter view, but the issue is not important here. Whatever one wants to say about the photon, let that be its "state."

11   This is not quite right: Leibniz famously argued against external relations. But he didn't convince anyone.

12   Bohm himself admitted that these sorts of difficulties beset any attempt to make his theory Lorentz invariant at a fundamental level. See, for example, Bohm and Hiley 1991.

8

Life in Elastic Space-time

Having confronted a formidable array of difficulties trying to adapt the quantum theory to Minkowski space-time we must now brace for a final shock: according to the *General* Theory of Relativity, physical space-time does not have a Minkowski structure. And having labored to produce Lorentz invariant theories which predict violations of Bell's inequality we must now come to grips with the idea that the actual universe admits of no Lorentz transformations. We will be able to do little more than touch on the further complications which accompany the shift from the Special to the General Theory, but we should get at least a broad overview of lay of the land.

The General Theory of Relativity (GTR) is the relativistic theory of gravitation. That Newtonian gravitational theory could not happily inhabit a relativistic space-time is obvious: Newtonian gravitational forces are determined by the *instantaneous* distribution of matter in the universe. We have already seen in the last chapter how this feature is central to making the Newtonian theory Galilean invariant. Having abandoned any univocal notion of instantaneousness, Relativity demands some deep revision of gravitational theory.

The mathematics needed for a rigorous treatment of the GTR overmatch any attempt at brief exposition. We will instead approach the theory by means of an analogy, a very accurate analogy which runs quite deep. Even without the use of any equations we can develop a clear picture of the theory, but we must begin by returning to the simpler realm of pure spatial geometry.

Quantum Non-Locality and Relativity: Metaphysical Intimations of Modern Physics, Third Edition. Tim Maudlin.
© 2011 Tim Maudlin. Published 2011 by Blackwell Publishing Ltd.

Non-Euclidean Geometry

The geometry we are most familiar with is the geometry of Euclidean space. In Euclidean geometry all triangles have interior angles equal to two right angles; straight lines meet at most once; and every direction in space determines a collection of parallel lines which never meet, keeping ever a constant distance from one another. However, Euclidean geometry turns out to be but one of an infinity of different possible geometries, all with their own unique properties.

The most obvious example of a non-Euclidean geometry is the intrinsic geometry of the sphere. On a sphere, the straightest and shortest paths between points are the Great Circles. Taking these circles as straight lines we are able to construct plane figures with strikingly non-Euclidean features. Thus, for example, consider a triangular circuit on the surface of the Earth which starts at the North pole, travels straight down to the equator, takes a right turn and follows the equator a quarter of the way around the Earth, then takes another right turn and goes back up to the North pole. This is a spherical triangle: each of the legs is part of a Great Circle. But each of the three angles is a right angle! This literally rectangular triangle cannot exist in a Euclidean plane, but lives happily in the non-Euclidean structure of the sphere.

The surface of a sphere is a model of a non-Euclidean plane geometry. Differences between spherical geometry and Euclidean geometry reflect differences in the intrinsic geometrical structure of the spaces being used. The exact mathematical expression of these differences requires the heavy machinery of differential geometry, but the leading point is obvious: the surface of the sphere has an intrinsic curvature to it which differentiates it from the Euclidean plane.

The surface of a roughly spherical object such as the Earth is clearly a curved surface in the three-dimensional space we live in, but the type of curvature relevant to non-Euclidean geometry is of another stripe. We say that the curvature of a surface is intrinsic since it does not depend on how that surface is embedded in a higher-dimensional space. Thus, for example, the existence of a rectangular triangle in a surface is proof that a space is non-Euclidean, whether or not it happens to be part of any higher-dimensional space. And in the other direction, being a curved surface as embedded in another space does not imply being intrinsically curved. A piece of paper rolled up into a cylinder retains its intrinsic Euclidean structure: one cannot draw any rectangular triangles on it.

Of the infinitude of possible geometries, then, Euclidean geometry is a single special case: the geometry of spaces with no intrinsic curvature. Spherical geometry is almost equally special, as it is the geometry of a space with a constant positive curvature. (Hyperbolic geometry is that of a space with constant negative curvature.) But most geometries are not so neat as these. Imagine, for example, doing geometry on the surface of the Earth as it really is. In some places, such as the Great Salt Flats, the geometry would very nearly be spherical (with such a small curvature as to seem Euclidean). But triangles drawn on the surface of the Alps would obey neither Euclidean nor spherical geometry. In such regions the intrinsic curvature fluctuates wildly from place to place, changing from positive to negative and back again. This intrinsic geometry admits no neat axiomatization; there are no general rules to assist the Alpine geometrician.

The full range of geometries includes the Alpine geometry and all others like it: spaces whose curvature varies from place to place in every possible fashion. Spaces like the surface of the earth that are smooth in some places and wrinkled in others; spaces which have Euclidean regions and spherical regions; spaces with pathological explosions of infinitely increasing curvature: all these can be described in the formalism of non-Euclidean geometry. Euclid explored just one sedate island in a raging sea of possibilities.

The General Theory

We can now state the fundamental analogy which reveals the heart of the GTR:

> As the general non-Euclidean geometries are to Euclidean geometry so the space-times of the General Theory are to Minkowski space-time.

Just as Euclidean geometry posits a single uncurved space, so the Special Theory posits a single uncurved space-time. The existence of a host of curved spaces points to the theoretical possibility of curved space-times. The heart of the General Theory is the claim that gravity, rather than being a force, is instead a manifestation of the intrinsic curvature of space-time. What appears in classical physics as the existence of a gravitational field is understood in the GTR as a deviation from Minkowski space-time structure.

How is the GTR able to trade off a classical force for something so seemingly unrelated as space-time structure? Again an analogy will help. Suppose Sam and Sue are roller-skating at the equator. Sam stands 4 meters north of the equator and Sue 4 meters south. They both push off heading due east and coast (figure 8.1). Being trained in Euclidean geometry, Sam and Sue will reason that since they are traveling on parallel paths, they have no fear of collision. If no forces act on them to deflect their trajectories, they should just coast for ever, keeping a constant 8 meters apart. But after a

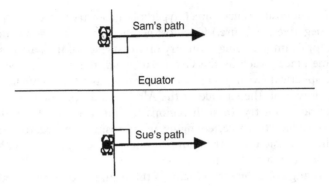

Figure 8.1   Roller-skating at the Equator

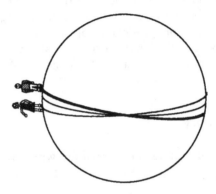

Figure 8.2   Sam and Sue from Above

while, Sam and Sue notice that they are slowly drifting together. They try to adjust their paths to avoid collision, but consistently find that they are drawn closer and closer. They try to keep a constant 8 meters apart by holding a stick between them, but find that the stick produces a constant pressure pointing away from the equator. They have to be pushed apart to keep from coming together. Eventually Sam and Sue conclude that there is some mysterious Force of Attraction drawing them to the equator.

From our vantage point high above in the space shuttle we can see just what is going on (figure 8.2). With no forces on them, Sam and Sue are each traveling along a perfectly straight path – in the intrinsic geometry of the Earth. They naturally will follow the Great Circle paths shown above, paths which each approach the equator. When holding the stick between them they follow latitude lines which are *not* straight lines in the intrinsic geometry: hence the stick must provide a force to deflect them off their natural paths. (It is easy to see that latitude lines are not straight by considering ones near the poles, where they are tight circles.)

Back down on Earth, Sam and Sue have decided to pursue their experiment. They bring along fat Uncle Sid, who weighs more than both of them put together. As Sid glides behind Sue, he follows just the path she does. Since they are attributing their "attraction" to a force, and since Sid needs a greater force to accelerate his greater mass, Sam and Sue conclude that more massive objects are more strongly attracted by their mysterious force than less massive ones. Indeed, to account for the fact that *all* objects follow the same paths, they eventually posit that their attractive force is proportional to the mass of the object attracted.

Sam and Sue are coming dangerously close to inventing a force that looks like Newtonian gravitation: a universal force that is proportional to mass. But they have been misled. There is no mysterious force of attraction operating at the equator, only a non-Euclidean geometry. Force-free objects are doing just what they should do: following straight paths.

According to the GTR, Newton made just this mistake. The so-called force of gravity is no force at all. Rather than producing forces, masses warp the intrinsic geometry of space-time from the Minkowski structure. As you sit on your chair you cannot feel the force of gravity pulling you down. Rather, you feel the surface of the chair pushing you *up*, as a little attention reveals. The chair keeps pushing you just as the stick pushes Sam and Sue, deflecting you off a straight trajectory through (not space but) space-time. The mass of the Earth has so deformed the geometry of space-time that a straight path from the point you occupy now would lead toward the center of the Earth. A particle with no forces on it would follow that straight path, obeying Newton's First Law. The cup you knock off the table follows that path. But your chair is continually accelerating you, that is, deflecting you off your natural trajectory. You feel this as the chair pushing up from below.

The exact details of how the distribution of mass and energy determines the intrinsic structure of space-time need not detain us. The fundamental analogy tells us enough about the space-times of the GTR to pursue our inquiry into non-locality.

Superluminal Constraints and the GTR

If one interprets the Special Theory as forbidding superluminal transport of something, whether it be mass or signals or causal connections or information, then the transition to the GTR is quite straightforward. Even in a General Relativistic space-time there are light cones defined at every point. The light cones may not "stack up" nicely as they do in Minkowsky space-time, they may tilt relative to one another (indeed, gravity is the tilting of light cones toward a massive body), but they still exist. So any prohibition against superluminal transport is easily transferred to this new milieu: one simply demands that the trajectory of the matter or signal or information stay within the light cone at every point. Violations of Bell's inequality at space-like separation has just the same consequences here as in Special Relativity (STR): there must be superluminal transmission of information and superluminal causal links. The extra complications of the GTR threaten no additional disasters, and hold out no additional hopes.

Lorentz Invariance and the GTR

If one interprets the fundamental relativistic constraint to be Lorentz invariance, the situation is much more dire. It is easiest to see why by returning to our analogy. Lorentz invariance demands that certain laws or quantities remain unchanged when one transforms from one Lorentz frame to another. The Lorentz frames are the global orthogonal rectilinear coordinate systems in Minkowski space-time. So Lorentz invariance demands that certain quantities remain the same when expressed in terms of any orthogonal rectilinear coordinate system. The global orthogonal rectilinear coordinate systems over the Euclidean plane are the Cartesian coordinate systems. So the analog to Lorentz invariance is the demand that some quantity or equation remain the same when expressed in terms of any Cartesian coordinate system.

When one passes from Euclidean to non-Euclidean geometry this demand for "Cartesian invariance" becomes meaningless. Global orthogonal rectilinear coordinate systems simply do not exist in most non-Euclidean spaces. One cannot, for example, lay down a Cartesian grid over the surface of a sphere. One of the difficulties is illustrated by the predicament of Sam and Sue. If one regards their paths as coordinate curves then either curves which should be parallel will intersect or, if the paths maintain a constant distance from one another, they will not be straight lines in the space. The evident distortions on maps of the globe employing the Mercator projection also illustrate how the sphere resists translation into the Euclidean plane. So in the generic case any mention of Cartesian coordinates is empty since such coordinates do not exist.

Following our analogy back to space-time, it is equally true that in a typical General Relativistic space-time there are no global orthogonal rectilinear coordinate systems. Since the Lorentz transformations express relations between such systems *in the GTR there are no global Lorentz transformations*. To require Lorentz invariance in the GTR is to make a meaningless demand.

Since Lorentz invariance has been touted as the conceptual core of Relativity, it comes as something of a shock to find the notion evaporate in the GTR. We must endeavor to understand, first, why the absence of

Lorentz transformations in the GTR has not caused great consternation and second, what formal properties guarantee conformity with the General Relativistic account of space-time structure in the same way that Lorentz invariance guarantees conformity with Minkowski space-time.

The lack of global Lorentz transformations is of little consequence for many theories because they are *local*. Theories such as relativistic electrodynamics postulate that the only data relevant to what happens at a point in space-time are data in the back light cone of the point, and further that the relevant physical quantities evolve continuously over space-time. In this case one can calculate the physical state at a point merely by knowing the physical state of a small region of space-time immediately preceding the point. One can, as it were, tear out a small patch of space-time and allow the theory to work in it alone.

The advantage of a local theory is that just as small patches of a non-Euclidean space such as the Earth's surface approximate Euclidean space (the Pythagorean theorem works fine in your back yard) so small patches of a General Relativistic space-time approximate Minkowski space-time. Gravitation warps space-time on a large scale; in the small scale, Special Relativity is nearly correct. The smaller the region under consideration the closer it approximates Minkowski space-time and the less relevant the GTR becomes.

In a local theory all fundamental physical transactions take place in infinitesimally small regions of space-time. It is enough for such theories to demand *local Lorentz invariance*. That is, one can require that the theory be Lorentz invariant when formulated in a flat space-time. Since the little local pieces of the space-time are essentially Minkowskian,[1] all of the physical transactions are consistent with the space-time structure. Global transformations between reference frames are otiose since the only relevant part of the transformation at any point of space-time is the transformation between coordinates in that point's immediate neighborhood.

But in a non-local theory local Lorentz invariance is evidently too weak. The local transformations among coordinate systems do not provide enough information to determine how the physical state of a region will be expressed in the new coordinates. Returning to figure 7.3, for example, we see that in the unprimed frame the right-hand photon at t = 3 is in a state of indefinite polarization. Suppose we wish to transform to the primed coordinates. No amount of purely local data will reveal that the same photon is in a state of definite polarization relative to the primed coordinates (and hyperplanes of simultaneity), nor what that polarization is. To know this we must determine that an observation is made on the left at t' = 3.2, and we must know the result of that observation. The exact nature of the coordinate transformation at places quite distant from the photon is essential in determining the physical state of that photon. Due to this disanalogy with local theories, local Lorentz invariance will not suffice to reconcile the quantum theory with General Relativistic space-time structure.

In order to approach the issue of consistency with the GTR, then, we need a formal condition which is neither Lorentz invariance nor local Lorentz invariance. We must be able to express in some precise way the fundamental ontological commitments of both the Special and General theories concerning space-time structure. The right way to express that commitment is as follows:

The only structure intrinsic to space-time itself is the metrical structure.

Putting the same point the other way around, if one must posit more intrinsic structure to space-time than is contained in the metric then one has abandoned the central posit of Relativity. This formulation applies equally well in the Special and General theories.

Let's see how this demand leads to Lorentz invariance in the Special Theory. In Minkowski space-time, the metric takes exactly the same form when expressed in terms of any orthogonal rectilinear coordinate system. That is, the metric itself is Lorentz invariant. So if a theory is also Lorentz invariant it cannot posit any more intrinsic structure than the metric itself. Since such a theory provides no means to distinguish one Lorentz frame over another, it cannot require more space-time structure than the metric, for the metric suffices to pick out the Lorentz frames.

Lorentz invariance in the Special Theory is therefore but a consequence of a deeper principle, the principle that space-time structure is nothing more than metrical structure. Any theory which posits, for example, an intrinsically privileged set of hyperplanes implicitly rejects Relativity. Such a theory cannot be Lorentz invariant (since the privileged set of hyperplanes is not), but the lack of invariance is merely symptomatic of the deeper problem. Since the metric *per se* privileges no hyperplane over another, the theory must appeal to some further spatio-temporal structure, structure which cannot be derived from the STR.

In the GTR, theories can respect the demand that the only intrinsic structure to space-time be the metric even though no Lorentz transformations may exist. Consistency with the GTR is then a matter of formulating a theory so that it employs nothing more than the metric when describing space-time itself.

The foregoing proposal must be carefully distinguished from a distinct claim which has bedeviled discussion of the GTR from the beginning. It is sometimes said that just as Lorentz invariance is the mark of a Special relativistic theory, so *general covariance* is the mark of a General relativistic theory. Or equivalently, that while the Special theory achieves a limited degree of generality by making all Lorentz frames physically equivalent, the General theory makes *all* coordinate systems equally valid. The suggestion that general covariance is the GTR's version of Lorentz invariance is completely wrongheaded. To see why, we must discuss general covariance.

Some coordinate systems are particularly convenient for expressing physical laws. A Lorentz frame, for example, is rectilinear: the trajectories of spatial coordinates are straight lines in space-time. Indeed it is precisely this feature which defines the inertial frames in both classical and Special relativistic physics. Inertial frames are convenient because they provide a simple correlation between acceleration and changes in coordinates: the spatial coordinates of an accelerating particle are not a linear function of time. Acceleration of a particle in an inertial frame can be expressed as the second time derivative of its spatial coordinates.

Other coordinate systems are not so convenient. We may, for example, set up a coordinate frame attached to a moving car, but in this case acceleration will no longer be a simply function of changes in coordinates. An object may well be at rest with respect to the car yet still be accelerating – if the car is. Trying to formulate standard Newtonian mechanics in such a frame demands the introduction of so-called fictitious forces which compensate for the fact that acceleration may not correspond to change in coordinates in the new frame. By the use of such frame-dependent fictitious forces Newtonian mechanics may be adapted to any reference frame at all.

General covariance just demands that a theory be formulated in such a way that it can be adapted to any given coordinate system, inertial or noninertial. This demand has no intrinsic connection to the GTR. As just noted, Newtonian mechanics can be given a generally covariant formulation, as can essentially any other physical theory. General covariance does not imply that all coordinate systems are "equal" or "the same": there is still a distinction between inertial and non-inertial systems, between accelerated and unaccelerated trajectories. The only reason that general covariance is particularly important for the GTR is that in some space-times there are no global inertial frames of reference. Any theory fit for General relativistic space-times must therefore not presuppose the existence of such frames and must be formulable in more arbitrary coordinate systems (such as the system of figure 2.1).

Compatibility of a theory with the GTR, then, amounts to this: the theory can require of space-time no more intrinsic structure than can be defined from the General relativistic metric. In particular, any structure which is to substitute for a classical notion of simultaneity must either be determined by the metric itself or must be shown to be definable from matter fields in the space-time.

Quantum Theories in Non-Minkowski Space-times

The various sorts of theories discussed in the last chapter must undergo different modifications if they are to be adapted to the GTR. Let's consider each in turn.

Theories with explicit backwards causation, such as Cramer's theory, have essentially the same explanatory resources available to them. This is because such theories have no need for any direct physical interaction between events at space-like separation. All of the interesting physics remains within the light cones; it is just that the past light cones are used on an equal footing with the future ones. Recall that in Cramer's theory correlations at space-like separation are enforced by a causal chain that connects each measurement event (as cause) with the production of the photon pair in its past light cone. The sorts of difficulties facing such a view remain basically unchanged, but may be magnified. We noted that we cannot think of the future events just sitting out there influencing present stochastic processes since the disposition of matter in the future may be in part determined by the results of those processes. Given the GTR the situation becomes even more extreme: the very spatiotemporal structure of the future may depend on the outcome of present stochastic processes. Influences from merely possible futures would have to propagate back through merely possible space-times, a notion which boggles the mind.

The many-minds theory is also unchanged, since it is a local theory anyway. There is a deep problem which confronts it, but it is a problem facing all attempts to marry quantum theory and General Relativity: understanding how the indeterminacy of quantum theory is to be reconciled with the seemingly definite structure of space-time. For example, consider again Schrödinger's cat. If the wave-function does not collapse (as it does not in the many-minds theory) then the cat ends up in a physical state in which it has no determinate location. One part of the wave-function corresponds to the cat standing up, say, and another to the cat lying down. But the cat standing up will have a different gravitational effect than the cat lying down. The space-time would have to go into a state of indefinite metrical structure. But a definite metric was a presupposition of the original quantum state: it is not clear how to superpose wave-functions which are defined over different space-times (and therefore different configuration spaces). Speculations on this topic have been put forward (see, for example, Penrose 1986; Károlyházy *et al.* 1986), but this puzzle takes us into the realm of quantizing Relativity, not relativizing quantum theory.

Theories incorporating hyperplane dependence also come through with their old interpretive problems magnified. In Minkowski space-time, such theories could be content with relativizing physical states to flat space-like

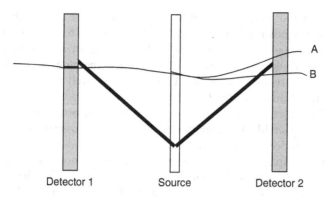

A
B

Detector 1   Source   Detector 2

Figure 8.3   Two Space-like Hyperplanes

hyperplanes. Since these hyperplanes are just the hyperplanes of simultaneity in the various Lorentz frames this choice seems quite natural. But in the curved space-times of the GTR there may well be no flat space-like hyperplanes: a curved space-time may not admit of a flat submanifold.[2] The only obvious response to this worry is abandoning the demand for flatness. In the space-times of the GTR, hyperplane dependence must become dependence on *any* space-like hyperplane.

The ontological profligacy of the Special Relativistic version of the theory expands geometrically in the GTR. Most surprisingly, the theory must now postulate that the physical state of a particle may differ relative to two hyperplanes *even though the hyperplanes coincide where they intersect the particle*. Figure 8.3 illustrates the situation: relative to hyperplane A the left-hand photon has a determinate polarization but relative to hyperplane B is does not. Both A and B are space-like hyperplanes, and there may be nothing in the metric of the space-time that allows one to insist that one rather than the other is to be preferred.

This modification of the hyperplane dependent theory does foreclose any suggestion that events on one side of the experiment somehow privilege a hyperplane, so that there is only one which gives the "real" state of the particle. As is obvious from figure 8.3, nothing happening on the left can account for a preference of hyperplane A over hyperplane B.

Bohm's theory and any theory which (unlike hyperplane dependent theory) postulates a single unique collapse of the wave-function are in serious difficulties in the GTR. The problem for such theories is that they demand something very like the absolute simultaneity of classical physics. In the STR, Minkowski space-time contains an embarrassment of riches: for every Lorentz frame there is a definition of simultaneity which has the formal features of classical simultaneity. So the problem there is to choose among all of the possible sets of simultaneity slices when neither the metric nor the distribution of matter may privilege any frame. One can cut the Gordian knot by simply positing a preferred frame, but this is to abandon the spirit of the Special theory. Note that one violates the spirit of Relativity not by rejecting the Minkowski metric but by adding something yet further to the intrinsic structure of space-time. This comports well with our proposal about compatibility with the GTR.

In the GTR these theories sometimes face the same problem: there may be many sets of space-like hyperplanes that could serve to stand in for classical simultaneity. Indeed, there may be many more candidate foliations of the space-time into space-like hypersurfaces since one cannot in general demand that the hypersurfaces be flat, for the reason cited above. But there may also be a new problem: in some models of the GTR there are no analogs of classical simultaneity at all.

To understand this we must introduce the notion of a *global time function*. Such a function is an assignment of numbers to the points in the space-time so that for every event in the space-time the numbers assigned to the points in its back light cone are less than the number assigned to it. The intuition is that every observer should agree that the points in the back light cone of an event precede it in time. (Equivalently, a time function ought to increase as one travels forward along a time-like trajectory.) A global time function permits definition of a kind of simultaneity: all of the points assigned the same number can be regarded as occurring "at the same time."

Classical space-times admit of only one global time function (up to arbitrary scaling parameters). It is in this sense that classical physics employs absolute time. Minkowski space-time admits of many global time functions, the t-coordinates of the Lorentz frames being among them. But some General relativistic space-times *admit of no global time function at all*. In these space-times, theories which demand a substitute for classical simultaneity cannot be saved even by the addition of more structure to the space-time. The problem is no longer an embarrassment of riches but destitution.

Models of the GTR which have no global time function are quite odd in many ways. The most famous of them is a space-time discovered by Kurt Gödel (see Hawking and Ellis 1973, pp. 168–70). The structure of this space-time is too complicated to present here; suffice it to say that Hawking and Ellis regard the absence of a global time function bizarre enough to brand the model as "not very physical" (1973, p. 170). We have come to the brink of a difficult argument about whether acceptance of the GTR entails the acceptance of even its most pathological models as physically possible. Not being prepared to take up that discussion, we must be content to note that any attempt on the part of Bohmians or of classical collapse theorists to retain any part of the GTR will force this argument upon them.

The lack of a global time function is especially worrying in a non-local theory. Gödel's universe has its puzzles for everyone, but at least a local theory has the luxury of independently analyzing bits of the space-time. A small chunk of Gödel's universe approximates a small chunk of Minkowski space-time, so any local theory can explain a particular event by just considering its immediate surroundings and using some relativistic equations. These local theories may run into difficulties when they try to extend local solutions to global ones, but a lot of physics can be done without considering the extension.

A non-local theory, in contrast, may not survive even temporarily in the Gödel universe. If the physical state of a photon is only defined relative to an unbounded space-like hypersurface, then the photon cannot have any state in Gödel space-time since there are no such hypersurfaces. The problem does not arise in extending a partial solution consistently to the whole space-time but in getting a partial solution at all.

In sum, some models of the GTR are extremely hostile to any proxies for classical Absolute Time. Others are far too accommodating: the choices for global time functions run amok. Extending any classical theory which makes essential use of non-local simultaneity to these contexts is likely to be much thornier than transplanting it to Minkowski space-time.

The GTR to the Rescue?

The malleable space-time structure of the GTR has so far only amplified our problems in coming to terms with quantum non-locality. There is, however, a seductive potentiality about the idea that space-time structure might be variable, a suggestive hint that perhaps all of our difficulties can be resolved if we recognize some heretofore hidden space-time structure. This hint has been tentatively voiced in the literature and so deserves some attention.

From the beginning, our puzzlement about violations of the Bell inequalities has been fueled by the intuition that particles greatly separated in space must eventually become causally isolated from one another, so that measurements carried out on one can have no influence on the other. We have seen how tremendously difficult it is to reconcile this intuition with the data. But if space-time structure is extremely malleable then it might turn out that places which we believe to be very far apart are really quite close to each other. In the GTR this sort of possibility is illustrated by the idea of a *wormhole* in space-time. It will be easiest to explain this idea if we first return to purely spatial geometry, so let's again visit Sam and Sue living on what they take to be a Euclidean plane.

One fine day, as they wander about their plane, Sam and Sue discover a remarkable set of correlations between occurrences at (what they take

Figure 8.4   "Non-local" Effects Between Sam and Sue

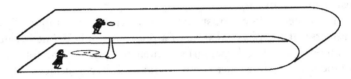

Figure 8.5   The True Spatial Structure

to be) extremely distant locations A and B. For example, Sam finds that when he whispers at A Sue can hear him at B, even though he cannot be heard at any intervening point (figure 8.4). After continued investigation they discover that nothing that takes place between A and B influences these distant correlations. Eventually they abandon their theoretical queasiness about the non-locality and embrace the idea of unmediated action-at-a-distance. Spatial propinquity is not, they conclude, required for direct causal connection.

One again, Sam and Sue have been duped, misled by mistaken presupposition about their space. In fact, the space they live in does not have a Euclidean structure at all. It will be convenient to visualize their two-dimensional space as embedded in a three-dimensional Euclidean space, keeping in mind that the embedding is only an inessential pedagogic technique. As can be seen in figure 8.5, although the majority of their space is Euclidean, it also contains a small tube connecting A and B, a so-called wormhole. Physical influences propagate continuously in the space. Causal influences are strong between A and B because A is very close to B. Sam and Sue, being unaware of this connection, think that all continuously propagating effects between the two spots would have to go the long way around. They might find that the whisper gets from A to B faster than light could – if the light takes the long route. But there is no non-locality, and the whisper might well travel at just Mach 1.

The possibility of wormholes in space-time arose quite apart from any consideration of violations of Bell's inequality. It has been suggested, for example, that a trip into a black hole would lead not to a singularity but to a wormhole. But once the idea of wormholes has been broached, it is hard to resist trying to take advantage of them to explain the distant correlations that arise in quantum theory. Indeed, the suggestion has been raised, albeit briefly, at least twice in print. Abner Shimony relates:

> I once suggested to [John] Wheeler that his idea of wormholes might be used to explain quantum mechanical nonlocality, for the path between two events through a wormhole might be time-like, whereas the path around the wormhole might be space-like. He was not encouraging. In fact, he said, "I am skeptical as hell." (Shimony 1990, p. 310)

Don Howard has also made the suggestion, somewhat more hopefully, although still as mere speculation. After describing a situation just like that of Sam and Sue, Howard makes the connection to quantum theory:

> The limit case, perhaps, corresponding to such phenomena as pair-creation, may be represented by a "thread" – an infinitely "thin" and sometimes infinitely "short" "pipette" – blossoming into a "tube." (Howard 1989, p. 252)

The vista here is so bracing – the furthest reaches of General Relativity explaining the deepest peculiarities of quantum theory – that it is perhaps cruel to shine our analytical light on it, but there are many reasons which support Wheeler's skepticism.

First, we should not let the fact that wormholes were first proposed as solutions to legitimate General relativistic problems fool us into thinking that the GTR itself suggests that creating a pair of photons could bring a wormhole into existence. General relativistic effects depend on masses, and where little mass density is involved one expects the space-time to approximate that of the STR. Recall that wormholes first arose in connection with black holes: regions of immense mass density.

Second, Howard's suggestion that wormholes be created along with pairs of correlated particles does not seem capable of explaining violations of Bell's inequality among particles not created together. But, as noted in the previous chapter, Yurke and Stoler (1992a,b) have argued that quantum theory predicts the strange correlations even among particles created apart. In any case, there is nothing in the nature of the relevant quantum state (the singlet state) that implies that the particles must have been created together.

Third, and more seriously, if one uses wormholes to explain the correlations then it becomes mysterious why "superluminal" signals cannot be sent, or why mass or energy cannot be transmitted between the wings of the experiment. A wormhole is a bit of space-time just like any other piece, and there would be no reason why commonplace particles and fields might not take the shortcut as well. If the wormhole construction were correct it would be very hard to see why there is any semblance of locality constraints at all.

Fourth, and most important, the postulation of wormholes really does nothing to alleviate the sorts of problems we discussed in the previous chapter.

Recall that the wormholes in question must connect bits of *space-time*, not of space. In this regard the example of Sam and Sue is misleading: their situation presupposes an absolute simultaneity, with the picture above being a snapshot of the structure of their space. But if we start putting wormholes into Minkowski space-time, we must determine which events are to be connected with which others. And once that connection has been made we will have picked out a preferred reference frame in the Minkowski space-time.

Figure 7.5 shows how any change in the potentialities of a photon due to the measurement of its partner would privilege a reference frame. In exact parallel, connecting the dots between the two ends of the worm-hole would pick out one hyperplane. If in chapter 7 we could not solve the problem of how the non-local connection is to be made, why think we will have any better luck finding a dynamics for our wormholes?

The GTR is no more hospitable to non-locality than the STR. Both theories work quite elegantly so long as nothing propagates faster than light. Neither theory explicitly rules out the possibility of superluminal processes. But discovering a truly relativistic theory that can deal with violations of Bell's inequality is an exceedingly difficult task, and the theories presently available entail such severe dislocations of our physical view that one must seriously consider whether our grounds for adhering to Relativity are really strong enough to justify such extreme measures.

Notes

1   Small patches of a curved space-time do not become identical to Minkowski space-time in the limit as the patches get small, but the deviations from Minkowski space-time appear only in quantities which are at least second order in the metric.
2   Indeed, some General Relativistic space-times admit of no global space-like hyperplanes at all, as we will see below.

9

Morals

"Yes, yes!" Walter interrupted this account. "First the four elements become several dozen, and in the end we're merely left floating around on correlations, on processes, on the dirty dish-water of processes and formulae, on something of which one doesn't know if it's a thing, a process, a phantom idea or a God-knows-what!"

Robert Musil

At the beginning of our inquiry we set ourselves some explicit questions. Since the Theory of Relativity has been variously interpreted as forbidding several sorts of superluminal propagation, we embarked on the project of relating the various constraints to the formalism of Relativity. We further inquired whether reliable violation of Bell's inequality by events at space-like separation could be reconciled with these constraints. For four of the proposed constraints I have argued that the results are unequivocal:

Violation of Bell's inequality does not require superluminal matter or energy transport.

Violation of Bell's inequality does not entail the possibility of superluminal signaling.

Violation of Bell's inequality does require superluminal causal connections.

Violation of Bell's inequality can be accomplished only if there is superluminal information transmission.

Our last topic, Lorentz invariance, yielded a tangle of unexpected proposals. Lorentz invariant theories which predict violations of the

Quantum Non-Locality and Relativity: Metaphysical Intimations of Modern Physics, Third Edition. Tim Maudlin.
© 2011 Tim Maudlin. Published 2011 by Blackwell Publishing Ltd.

inequality may be formulable if one admits either explicit backwards causation or hyperplane dependence. More radically, one could interpret the violations as indicating the existence of a single preferred reference frame, a frame undetectable by any empirical means. Even more radically, one could adopt the many-minds theory and deny that there are any violations of Bell's inequality by events at space-like separation: the relevant correlations exist only in individual minds. All of these options become yet more bizarre when one shifts from Special to General Relativity.

I have not advocated a single choice among these options. It is very difficult to imagine a clear set of uncontroversial standards of plausibility that would yield a definitive decision here. Perhaps other, more unusual and counterintuitive possibilities may be articulated. There is, however, one final moral to be drawn from the teratological collection we have assembled.

Choices among empirically indistinguishable theories are commonly made on aesthetic grounds. In particular, possibilities are often rejected on the grounds that had God made the world in accord with such a theory, he would have been malicious. For example, it is possible to design theories that are empirically equivalent to the Special Theory of Relativity but that posit Newtonian Absolute Space and Absolute Time. If one supposes that Maxwell's equations hold in only in the One True Reference Frame one can then derive that the behavior of electromagnetic clocks and measuring rods will not allow one to discover which inertial frame is the One True One (see Bell 1987, ch. 9). Rods will shrink and clocks slow down in just such a way that the speed of light *seems* to be the same in all frames, though it is not. Such a theory, although logically consistent and empirically impeccable, is generally considered to be inferior to Special Relativity. The grounds for this judgment are not usually made very explicit, but the general idea is that it would be awfully deceptive to create a world with Absolute Space and then use the laws of physics to hide its existence from us.

Proceeding in this way, partisans of any of the above-mentioned theories may attempt to defend their view by a *reductio*. They may correctly note that according to every one of the rival theories, God was malicious, and having thus eliminated every other possibility, claim their own theory the victor. The problem is that *every* partisan can argue in this way since *every* theory posits some funny business on the part of the Deity.

If there is explicit reverse causation, why do all known and controllable causal processes have causes which precede their effects? If physical states are so radically hyperplane dependent how has so much completely local physics, which attributes states which don't depend on hyperplanes, worked so well? If there are wormholes between points in space-time, why can correlations be enforced through them but not signals sent? If the world we experience is only in our mind why does the postulation of mind-independent determinate physical states work so well?

More generally, if there are real superluminal causal connections why are they so ephemeral? Why can't they be used to transmit signals, or be otherwise made more manifest? If the world is not, at base, relativistic, why does it so much seem to be? And if it *is* relativistic, so the space-time structure alone does not preferentially connect an event with any unique class of space-like separated events, then what determines which events at space-like separation interact with one another?

One way or another, God has played us a nasty trick. The voice of Nature has always been faint, but in this case it speaks in riddles and mumbles as well. Quantum theory and Relativity seem not to directly contradict one another, but neither can they be easily reconciled. Something has to give: either Relativity or some foundational element of our world-picture must be modified. Physicists may glory in the challenge of developing radically new theories in which non-locality and relativistic space-time structure can more happily co-exist. Metaphysicians may delight in the prospect of fundamentally new ontologies, and in the consequent testing and stretching of conceptual boundaries. But the real challenge falls to the theologians of physics, who must justify the ways of a Deity who is, if not evil, at least extremely mischievous.

10

New Discoveries and Deeper Insights: The View from 2010

The first edition of this book was published in 1994. The state of play at that time was in some respects perfectly clear and in others somewhat equivocal. What was clear – and still is – is that no local physical theory, in Bell's sense, can predict violations of Bell's inequality for experiments performed at space-like separation, if the choice of which experiment to perform is treated as a free variable when making predictions. And it is clear that refusal to treat the choice of experiment as a free variable implies some sort of hyper-conspiratorial determinism (since the experimental apparatus can be hooked up to whatever sort of randomizing device we like) that is incompatible with any serious sort of physical theory. So we must either abandon locality or abandon the predictions of quantum theory for events at space-like separation. I have sketched how some versions of the many-worlds interpretation of quantum theory appear to do the latter, and considered in some detail how locality might be abandoned in a technically precise way. The clearest and cleanest way to implement a non-local physics would be to introduce a preferred foliation into space-time, and use that foliation in the same way that absolute simultaneity is used in the non-relativistic version of the theory. This solution cuts the Gordian knot by rejecting the basic relativistic picture of space-time structure, and it remains an open and tempting option. The non-locality becomes physically robust and clear; the dependency of events in one region on events at space-like separation plainly calculable. This route leaves us desiring an explanation for the empirical inaccessibility of the foliation, either in principle or in practice, but that desire can be satisfied by the physics itself. What is and is not empirically accessible by experiment is itself a question for physics to answer, and we

Quantum Non-Locality and Relativity: Metaphysical Intimations of Modern Physics,
Third Edition. Tim Maudlin.
© 2011 Tim Maudlin. Published 2011 by Blackwell Publishing Ltd.

have no standing to demand that all physically real objects or structures be capable of being laid open to our view. This sort of approach would use physics to explain the *appearance* of a relativistic world as a physical consequence of a fundamentally non-relativistic physical reality. Although logical empiricists would object to this sort of analysis, it has become abundantly clear that no physics can or should be bound by such empiricist structures.

Nonetheless, a fundamental rejection of Relativity, in the form of a physical foliation, has few enthusiasts. Relativity is a very beautiful theory, and many would prefer not to abandon it even in the face of Bell's results. So the question remained whether a fully relativistic method of implementing some sort of non-locality could be developed. With respect to theories that invoke the collapse of the wave-function at a fundamental level, in 1994 I wrote "In sum, relativistic theories of wave collapse do not yet exist" (p. 204). No argument available at that time could rule out such a theory, but neither had any been invented. So things stood then.

When the second edition of this book was published in 2002, I wrote in the preface "Finally, I must note that although there has been some discussion of Bell's theorem and non-locality in the eight years between the two editions of this book, there has been, to my knowledge, no fundamental change in the basic logic of the situation, and no real progress in reconciling quantum theory and Relativity" (p. xi). The same cannot be said in 2010. Indeed, recent discoveries have been the impetus for a third edition. We now have in hand a fully relativistic theory that predicts violations of Bell's inequality for experiments at space-like separation. So one open question has been closed: relativity and non-locality are not logically incompatible. But, to use the metaphor from the last chapter, the Deity is still mischievous. The theory that achieves this feat is sufficiently peculiar in other ways that it remains reasonable to admit a fundamental foliation to space-time rather than accept the basic physical ontology of this new theory. But we are getting ahead of the story.

The GRW Theory

The theory that reconciles relativistic space-time structure with violation of Bell's inequality was published by Roderich Tumulka (Tumulka 2006). It is a variant of the spontaneous collapse theory developed by GianCarlo Ghirardi, Alberto Rimini, and Tulio Weber, universally known as the GRW theory (Ghirardi *et al.* 1986). Ghirardi, Rimini, and Weber were not particularly concerned with Relativity or Bell's theorem; rather, their concern was to provide a precise theory according to which the wave-function collapses. Standard quantum theory is not physically precise about wave-function collapse in several ways. First, if one asks the standard theory exactly *when* or *under what circumstances* the wave-function fails to evolve in accordance with a linear equation of motion, the answer is "when a measurement occurs." And if one asks the standard theory *how* the wave-function collapses, i.e. what the post-collapse state is, the answer is "one or another of the eigenstates of the operator that corresponds to the measurement, with the probabilities for different outcomes calculated in the standard way from the pre-collapse state." So the *when* and the *how* of collapse in the standard theory are both tied to the notion of measurement. To apply the theory to a physical situation, one has to know whether that situation contains any measurements at all, and, if so, what operator corresponds to the measurement.

But "measurement," as Bell insisted, is simply not a precise enough notion on which to found a physical theory. The sorts of experiments we call "measurements" are physical interactions, but there is no physical characteristic that distinguishes measurements from other sorts of interactions. As Bell writes, echoing a similar remark by Einstein,

> It would seem that the theory is exclusively concerned about 'results of measurement,' and has nothing to say about anything else. What exactly qualifies some physical systems to play the role of 'measurer'? Was the wavefunction of the world waiting to jump for thousands of millions of years until a single-celled living creature appeared? Or did it have to wait a little longer, for some better qualified system ... with a Ph.D.? If the theory is to apply to anything but highly idealized laboratory operations, are we not obliged to admit that more or less 'measurement-like' processes are going on more or less all the time, more or less everywhere? Do we not have jumping then all the time? (Bell 1990, pp. 19–20)

In short, there is nothing wrong with collapses *per se*, but a collapse theory needs to give an account of the collapses – both the *when* and the *how* – in straightforward physical terms, with no mention of "measurement."

The GRW theory achieves this feat for non-relativistic quantum mechanics. The general form of the theory is so simple and transparent that one cannot but be amazed that it took over half a century to discover. To the *when* question, the GRW theory answers: from time to time. More precisely, in the original version of the theory each fundamental particle has a fixed probability per unit time of suffering a GRW collapse or "hit." The probability for each particle is extremely low: a single particle will experience a collapse on average only once in 10^8 years. Over the course of all recorded human history, less than 1 percent of particles will have suffered "hits." The chance of any smallish collection of particles (less than a million) experiencing a hit over the course of a laboratory experiment is negligibly small. This would explain why experiments done on individual particles or small collections of atoms would not yield direct evidence of collapses.

The GRW theory solves the problem of specifying the circumstances of collapse by cutting the Gordian knot: there are no circumstances that can either promote or delay a collapse. Instead, one needs a new constant of nature that specifies the mean time between collapses. This constant is subject to some empirical constraints, but at present the value of 10^8 years is compatible with all observations.

As to the question of *how*, GRW asserts that the effect of a collapse is to localize the wave-function of the hit particle in space. Mathematically, this is achieved by multiplying the wave-function, expressed as a function of space, by a Gaussian (bell curve). Once again, a new constant of nature is required, specifying the shape of the Gaussian, and once again there is at present a range of values compatible with observation. The original GRW paper chooses a width of the Gaussian as about 10^{-5} cm, rather larger than the size of an atom but much smaller than a red blood cell. The main requirement is that the Gaussian be narrow enough to resolve any macroscopic ambiguity in the location of a particle. We will return to this topic directly, but let's first pause to consider what has been accomplished so far.

The source of the measurement problem in standard quantum theory is the appearance of the concept of measurement in the fundamental axioms of the theory. Those axioms specify the dynamics of the wave-function of a system. The GRW dynamics for the wave-function, in contrast, are fully specified without any mention, direct or oblique, of measurements. Most of the time, the wave-function of a small collection of particles evolves in accordance with the usual deterministic linear dynamics (Schrödinger's equation). The deviations from this evolution, the "hits," occur at random, with fixed probability per unit time. The locations of the hits are also at random, but they are more likely to be centered at locations where the squared amplitude of the wave-function is higher. The effect of a hit on the wave-function is always the same: multiplication by a Gaussian function that is rather narrow at macroscopic scale.

How does this change in the standard dynamics resolve the Schrödinger cat problem, especially since the chance of any particular particle in the cat being hit in a given minute, or hour, or day, is minuscule? One key to the resolution lies in the observation that the particles in the cat are highly entangled with one another. Suppose, for example, that we contrive the sort of situation that Schrödinger discussed: absent any collapse of the wave-function, the quantum state of the cat at some time will with certainty be:

$$1 / \sqrt{2} |\text{cat alive} > + 1 / \sqrt{2} |\text{cat dead} >$$

where |cat alive> is the wave-function of a cat that has not been poisoned and |cat dead> the wave-function of a cat that has been. The wave-function is evidently symmetric between the two cases, and so does not correspond either to a live cat or to a dead cat. This is the result that Schrödinger found intolerable since he accepted that any particular cat will either be dead (and not alive) or alive (and not dead) at the end of the experiment. The symmetry of the wave-function needs to be broken to resolve the situation one way or another, but the linear dynamical equation for the wave-function will not break the symmetry.

A GRW hit on a particle, however, will break the symmetry. Consider, for example, a single particle in the cat's heart at some moment. If the cat is alive, and the heart is beating, then the position of the particle in space will be X_{alive}, but if the cat is dead the position will be X_{dead}. In general, X_{alive} and X_{dead} will differ from each other by much more than 10^{-5} cm. So if that particular particle happens to suffer a GRW hit when the cat is in the state $1/\sqrt{2}|\text{cat alive}> +1/\sqrt{2}|\text{cat dead}>$, the Gaussian representing the effect of the hit will either be centered somewhere very near to X_{alive} or somewhere very near to X_{dead}, with 50 percent chance of each. Intuitively, the hit will "localize" the particle either near X_{alive} or X_{dead}. *And since the positions of the rest of the particles in the cat's heart are entangled with the position of this particle,* the hit on the single particle will equally break the symmetry for all the rest. A single GRW hit on any such particle in the cat will either yield a wave-function very close to |cat alive> or a wave-function very close to |cat dead>, with 50 percent chance of each.

It is at this point that the "macroscopic" character of the cat becomes important. For although it is overwhelmingly unlikely that *any particular* particle in the cat will suffer a hit in a given minute, it is overwhelmingly likely that *some particle or other* in the cat will suffer a hit in much, much less than a second. If the cat has 10^{23} particles (a low estimate), then we can expect a hit every 10^{-8} seconds, with any single hit enough to change the wave-function so it is either very close to |cat alive> or very close to |cat dead>. As Bell says, "Quite generally, any embarrassing macroscopic ambiguity in the usual theory is only momentary in the GRW theory. The cat is not both dead and alive for more than a split second" (Bell 1987, p. 204).

The GRW dynamics, then, seems to resolve Schrödinger's puzzle without invoking the notion of measurement: the wave-function of the cat either ends up very close to |cat alive> or very close to |cat dead>, with the appropriate chance of each. But there are still some problems to be resolved. One might wonder, for example, whether being "very close" to |cat alive> is really enough to assure the existence of a live cat: the GRW dynamics will not leave the cat's wave-function *exactly* in |cat alive> or *exactly* in |cat dead> even after a hit. This question has attracted some discussion (cf. Albert and Loewer (1996); P. Lewis (1995)). But Bell pointed out a much more troubling issue: so far, our entire discussion has been only about the dynamics of the wave-function. We have not considered at all what connection there might be between the behavior of the wave-function and *the disposition of matter in space-time*. In short, we have not been provided any space-time account of the cat at all. And since we describe cats – living and dead – in terms of the disposition of matter in space-time, it is not yet clear how to connect all of this talk of the wave-function to cats at all. Furthermore, without such an account of what there is in space-time, we cannot begin to address the question of how to adapt the GRW theory to a relativistic space-time.

Local Beables for GRW

It is easy to form the impression that quantum physics is only about wave-functions. The Schrödinger equation, for example, specifies how the wave-function of a system evolves, as does the GRW hit mechanism. But the wave-function of a system does not, by itself, specify any particular physical contents in space-time. Without such a specification, one is left at a loss about how to connect the wave-function, whatever it may be, to what we seem to know about the world. As Bohr insisted, our description of laboratory operations and their outcomes, or more generally our description of whatever we take to be observable facts that can serve as evidence for or against a theory, is couched in a "classical" language. That is not to say that we need to use classical *physics* or classical *mechanics* in describing such evidence. The issue is more basic than that: our description of observable behaviors of things is given in terms of how they behave in space and time. We might say, for example, that at the end of an experiment a pointer on an apparatus ended up pointing to the right rather than to the left. *In order for such a state of affairs to have any relevance to a proposed physical theory, that physical theory must provide an account of conditions under which the pointer points one way rather than another.* But on the face of it, this is a claim about how matter ends up being arranged in space-time. So the obvious way for the physical theory to account for this result is for it to postulate the existence of something-or-other in space-time, and to provide predictions (either deterministic or probabilistic) about how that something-or-other will behave.

But wave-functions do not exist in space-time; they do not have values at particular space-time points. Bell again:

> It makes no sense to ask for the amplitude or the phase or whatever of the wavefunction at a point of ordinary space. It has neither amplitude nor phase nor anything else until a multitude of points in ordinary space are specified. (Bell 1987, p. 204)

Mathematically, the wave-function is defined on the *configuration space* of a system, an abstract mathematical space in which every point represents a hypothetical configuration of the system of interest. That is, specifying a single point in the configuration space is tantamount to specifying exactly where, in ordinary space, every particle in the system is. If we want a physics in which things happen in space-time, then we need a physics in which more exists than the wave-function.

When we say that a GRW hit tends to localize the wave-function of a particle in space, what we properly mean is something more subtle. The wave-function is defined on the configuration space of a system, and three dimensions of that high-dimensional configuration space are treated as representing "the position of particle p." Of course, just what this means if there is no particle p that always has a definite position is not clear, but that does not prevent this sort of language being used all the time. The mathematical effect of a GRW hit on the wave-function is to multiply the wave-function by a three-dimensional Gaussian in the three dimensions associated with the particle. And the effect of this is to radically reduce the amplitude of the wave-function at any point in configuration space which, as it were, corresponds to a configuration in which particle p is far from the center of the Gaussian.

This effect on the wave-function is commonly called "localizing the particle," but the exact physical meaning of this phrase is not clear. Using the standard terminology, one is tempted to say that the probability of *finding* particle p near the center of the Gaussian is now quite high, but this sort of phraseology is clearly question-begging if by "finding" one means to refer to the outcome of a "position measurement." Our whole project, recall, is to make sense of the physical account given by quantum theory without making reference to measurements. One is also tempted to say that after the collapse the probability of particle p *being* near the center of the Gaussian is quite high. But to make good on such a claim one must postulate the existence of particle p as an entity with a precise position, even when the wave-function is not in an eigenstate of the "position operator." Most physicists are reluctant to do this.

Rather, most physicists tend to think of the particle as having only a "fuzzy" position, with "more of it" located in places that correspond to its location in configurations which are assigned high amplitudes by the wave-function. This suggests a picture in which particle p is "smeared out" in space, and the effect of the GRW hit is to concentrate most of the smear within 10^{-5} cm of a particular location. We can make precise sense of such a picture, but only by being explicit about it. If we want to say that something or other *exists* in space-time, then we ought to say exactly what it is and how it is to be described.

Bell introduced a useful piece of terminology here: *local beables*. In his paper "The Theory of Local Beables," Bell began by coining the term "beable" for what philosophers would call the *ontology* of a theory: an inventory of what the theory postulates as existing. In any standard interpretation of quantum theory, the wave-function or quantum state of a system will be in this inventory: it is a feature of physical reality. But the wave-function, as we have seen, is not a magnitude defined on physical space or space-time. Bell reserved the term "local beable" for these:

> We will be particularly concerned with the *local* beables, those which (unlike, for example, the total energy) can be assigned to some bounded space-time region. For example, in Maxwell's theory the beables local to a given region are just the fields **E** and **H**, in that region, and all functionals thereof. ... Of course, we may be obliged to develop theories in which there *are* no strictly local beables. That possibility will not be considered here. (Bell 1987, p. 53)

The advantage to having local beables in a theory is that it becomes easy to understand how the predictions of the theory connect to the description of what is observed in a laboratory or in nature. The pointer on an apparatus moving to the right, for example, should just be a matter of some local beables postulated by the theory coming to occupy a particular region of space-time. Evidently, only local beables, things that exist in some bounded region of space-time, can do this.

An obvious possibility arises here: since the wave-function is defined on the configuration space of a system, one might suggest that the system must always *have* a configuration, i.e. that the particles in a system must always *have* some particular, exact, determinate position. Indeed, one might suppose that this is already implied by the use of the term "particle": particles are physical items that are always in some definite location. The task of physics is then to specify how these particles can or must move in particular situations. Such laws of motion might yield, for example, that the particles that make up a pointer will have all moved to the right by the end of an experiment, so the pointer will point to the right. We can call such a physical picture a *particle ontology*. As we saw in appendix B, a non-relativistic quantum theory with a particle ontology exists: Bohmian mechanics. In that theory the wave-function never collapses. The GRW theory of wave-function dynamics, though, has never been allied with a particle ontology. The original paper of Ghirardi, Rimini, and Weber does not address the question of local beables at all: the focus is entirely on the wave-function (or, more precisely, the density matrix) of a system. Such disregard of local beables was standard practice in physics at the time, and remains standard practice today. So if we want to discuss local beables in the context of GRW we have to supplement the dynamics of the wave-function with an account of the local beables.

It is easy to fall into the error of thinking that the GRW theory already contains an ontology of local beables, and indeed that any theory with a wave-function already contains some local beables. Presentations of non-relativistic quantum mechanics tend to employ the term "particle" even while they deny that the theory postulates items that always have definite locations in space, i.e. particles. And if one asks after the location of these "particles," the standard theory will invoke a "position observable" that is used to make predictions about the outcome of a "position measurement." Again, this terminology cannot be taken very seriously: if the "particle" does not *have* a definite position at a given time, then no interaction can *reveal* such a position. The interaction cannot properly be said, in this sense, to be a measurement. Nonetheless, there are experimental arrangements called "position measurements," and the theory provides probabilities for these "position measurements" to yield different possible outcomes. We may call such an outcome "finding the particle in a particular location" bearing in mind that the theory does not assert that the particle *was* in that location before the "position measurement" was carried out.

If one asks after the "location" of a "particle" in this circumstance, it is tempting to say that the particle has an "indefinite location" or is "smeared out," reflecting the fact that a "position measurement" could have many different outcomes. And from here it is but a short step to say that the particle is somehow more like a classical field or a classical wave than a classical particle: it is present in many different locations, more of it in one place and less of it in another. That is, it is easy to treat the *probability density* for "finding" the "particle" somewhere by a "position measurement" as if it were a *matter density* of the "particle" itself. Where the probability is higher, more of the "particle"'s matter exists; where it is lower, less does. Taking this picture seriously yields a *matter density ontology*, as opposed to a particle ontology.

The difference between a particle ontology and a matter density ontology is easy to illustrate. Consider the classic two-slit experiment: we fire a series of "particles" through a wall with two apertures toward a fluorescent screen beyond. A single spot eventually forms on the screen for each "particle" sent, and the collection of spots displays the alternating bands characteristic of interference. What exactly happens in the space between the "particle" source and the screen? What are the local beables in that region during the firing of a single "particle"?

According to the particle ontology, each firing really does involve a single particle with a definite location, and that particle follows some continuous trajectory from the source to the screen. Each particle goes through exactly one of the apertures. Since the interference bands emerge only when both apertures are open, there must be something physical that is sensitive to both apertures, but in this picture that physical item is not directly the particle itself: it is the wave-function of the particle. The wave-function evolves differently when both apertures are open than it does when only one is. According to Bohmian mechanics, the wave-function guides the particle, which explains

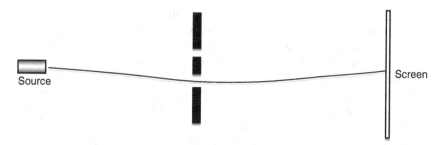

Figure 10.1   Time-lapse Diagram of Local Beables in Bohmian Mechanics

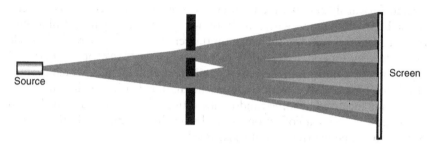

Figure 10.2   Time-lapse Diagram of Local Beables in GRW with Matter Density Ontology

how the particle trajectories depend on the state of both apertures. Still, each individual particle goes through exactly one slit (figure 10.1).

The spot on the screen forms where the particle strikes it, so the interaction with the screen really does constitute a measurement that reveals a pre-existent physical fact.

The matter density ontology presents a completely different account of things. Since the quantum-mechanical probability for "finding" the "particle" at various locations in the device is spread throughout the space, including being non-zero at each of the two apertures, in this theory the "particle" itself spreads out in space, with some of it going through each aperture. Once again, it is really the wave-function (not the matter density) that gives rise to the interference bands, but in the GRW picture the individual spots on the screen do not reflect where a "particle" was just before the spot forms. Rather, the position of each individual spot is determined by a spontaneous collapse, a GRW hit that takes place *after* the "particle" has interacted with the screen. According to the matter density ontology, the matter density of each individual particle arrives at the screen spread out, with interference bands in place (figure 10.2).

According to the GRW dynamics, when a collapse finally occurs (and one becomes more likely as more and more of the "particles" in the screen get entangled with the "particle" sent), it is more likely to be centered where the mass density is larger. The effect of the collapse is to localize the matter density, to concentrate almost all of it within 10^{-5} cm of the collapse center. So after many "particles" have been sent (each with a matter density that looks just like figure 10.2) and many collapses have occurred (all being centered at different locations, but with probabilities proportional to the matter density in a region), the usual interference bands will build up – as a straightforward accretion of matter – on the screen.

The evident differences between figure 10.1 and figure 10.2 illustrate how distinct these two sorts of ontology are. These are, at the most fundamental level, different physical accounts of the world. Yet no experiment could straightforwardly reveal which – if either – of the two accounts is correct. In the Bohmian picture, when we "look for" a particle (by means of a fluorescent screen) we find it: the mark on the screen records where the particle was before we looked. In the GRW picture, when we "look for" a "particle" (by means of a fluorescent screen) we initiate an interaction, which eventually leads to a collapse, which causes the matter density of the "particle" to become localized at some random location. Yet in both cases, at the end of the day, we have a collection of marks on the screen, and both theories predict the same distribution of those marks.

At this point, one might well wonder what real work the local beables are doing at all. If we can't empirically determine even whether the local beables are distributed as depicted in figure 10.1 or as in figure 10.2, why bother postulating them in the first place? They seem to have no empirical significance, and hence no scientific significance. They are merely the playthings of philosophers. They are truly "hidden variables."

The problem with this suggestion is what happens if we carry it out in a thoroughgoing way. Let's draw a third picture of our experiment on the assumption that there are no local beables at all (figure 10.3).

The problem, of course, is that we can't just eliminate the local beables for the "particle" being shot at the screen. The particles – electrons, protons,

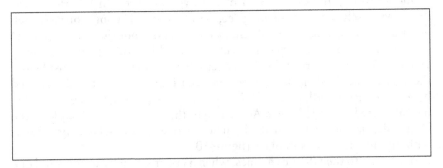

Figure 10.3 Two-slit Experiment with No Local Beables

neutrons, and so on – are also the constituents of the source, and of the wall, and of the screen. If there are no local beables associated with any particles, then there are no local beables full stop, and nothing at all takes place in any region of space-time. But this is flatly inconsistent with the description of the experimental conditions: they are given in terms of distances from source to the wall and from the wall to the screen, the size and location of the apertures, and so on. It is these very data that allow us to make a prediction for the outcome of the experiment in the first place. If there are no local beables at all, then there is nothing for this space-time description of the experimental set-up to describe.

It is often said that the most profound problem with the Copenhagen interpretation of quantum theory is the measurement problem: the theory invokes measurement in its fundamental axioms without a physical account of what a measurement is. But if one has Bohr's explicit views in mind, there is an even more profound problem than this. Bohr insisted that the description of an experimental situation and its outcome be given in "classical" language: the language of material objects arrayed in space-time. Yet he also insisted that there could be no space-time account at all of the microscopic realm. But it is an unshakable postulate of physics that the macroscopic objects mentioned in the description of the laboratory are, in fact, just collections of microscopic objects: electron, quarks, and so on. So if the latter have no space-time properties, neither do the former, and the "classical" language that Bohr insists on has nothing to describe.

This conceptual flaw in the Copenhagen account was pinpointed (of course) by Bell:

> The kinematics of the world, in this orthodox picture, is given by a wavefunction (maybe more than one?) for the quantum part, and classical variables – variables which *have* values – for the classical part: $((t, q, ...), X(t),)$. The X's are somehow macroscopic. This is not spelled out very explicitly. The dynamics is not very precisely formulated either. It includes a Schrödinger equation for the quantum part, and some sort of classical mechanics for the classical part, and 'collapse' recipes for the interaction.
>
> It seems to me that the only hope of precision for the (Ψ, x) kinematics is to omit completely the shifty split, and let both Ψ and x refer to the world as a whole. Then the x's must not be confined to some vague macroscopic scale, but must extend to all scales. In the picture of deBroglie and Bohm, every particle is attributed a position $x(t)$. Then instrument pointers, assemblies of particles, *have* positions and experiments *have* results. (Bell 1990, p. 30)

The only way to make sense of the sorts of space-time diagrams we have been concerned with is via the postulation of at least some local beables. Without them, none of our questions (concerning, for example, experiments performed at space-like separation) can even be stated. The particle ontology and the matter density ontology provide examples of how local beables can be explicitly introduced into a quantum theory.

Our discussion of GRW, then, requires a commitment to some local beables or other. As it turns out, to adapt the GRW theory to Minkowski spacetime, a completely different set of local beables seems to be required. These too were originally suggested by Bell.

The Flash Ontology

In his early explication of the GRW theory, Bell confronted the question of local beables. The comment cited above about the wave-function not having any amplitude or phase at a point in physical space was followed by this suggestion:

> However, the GRW jumps (which are part of the wavefunction, not something else) are well localized in ordinary space. Indeed each is centered on a particular spacetime point (x, t). So we can propose these events as the basis of the 'local beables' of the theory. These are the mathematical counterparts in the theory to real events in definite places and times in the real world (as distinct from the many purely mathematical constructions that occur in the working out of physical theories, as distinct from things that may be real but not localized, and as distinct from the 'observables' of other formulations of quantum mechanics, for which we have no use here). A piece of matter is then a galaxy of such events. (Bell 1987, p. 205)

Bell's suggestion is both clear and extremely disquieting. We must take it to heart before we can proceed.

We have already noted that the GRW hits are quite sparse for individual "particles": most "particles" will not suffer a hit in the entirety of human history. But according to Bell's proposed set of local beables – since denominated the *flash ontology* – the only time anything associated with the "particle" exists in space-time is when there is such a hit, such a collapse of the wave-function. *So according to the flash ontology, most of the "particles" in your body will leave no mark at all in space-time throughout your lifetime.* The collection of local beables in this theory is very sparse indeed.

Here is another way to appreciate the sparseness. Figures 10.1 and 10.2 are time-lapse representation of the local beables ontology during the two-slit experiment according to the particle ontology and the matter density ontology. We have used the time-lapse to show the complete trajectory of the particle (or "particle") from source to screen. But if we took an instantaneous snapshot of the local beables at some moment, the diagrams would not look very different. In figure 10.1, the particle being shot would show up as just a single point somewhere along the indicated trajectory, while the source,

Figure 10.4   Time-lapse Diagram of the Flash Ontology

wall, and screen would be unchanged. In figure 10.2, such a snapshot would reveal a thin vertical slice of the mass density of the "particle," and again the source, wall, and screen would be unchanged. But a snapshot of the experiment according to the flash ontology would almost always look like figure 10.3: there would be no local beables at all. Since the flashes occur only when a collapse occurs, and since at most times no collapse is occurring, at most times space-time is completely empty.

How is this any better than having no local beables at all? If we return to our time-lapse picture, and show all of the flashes that occur during, e.g., 1 second, the result is quite different. During that second the "particles" in the source and in the wall and in the screen will suffer millions of hits, and that collection of hits will reveal a familiar space-time world (figure 10.4).

We have added a physical pointer containing many particles to the picture, and imagine that detectors in the screen are hooked up to the pointer so it slides to point where the detector fires. After each "particle" goes through the device the pointer ends up pointing somewhere, and if we collect together all of those positions as data the usual interference bands emerge. But note one curious fact: in the region between the source and the wall and the wall and the screen *nothing at all exists.* Save in the very unlikely event that the "particle" suffers a hit in flight, it will leave no trace in space-time. Still, the data produced by the pointing arrow will be exactly what quantum theory predicts.

Although the time-lapse distribution of local beables in macroscopic objects (such as the wall) looks the same in figures 10.1, 10.2, and 10.4 at the scale of these diagrams, they would be completely different at microscopic scale. Consider a small patch of space-time in the wall, extended in both space and time. According to the particle ontology it will be traversed by particle world-lines; according to the matter density ontology it will be filled with different densities of matter roughly 10^{-5} cm wide; and according to the flash ontology it will contain a comparatively

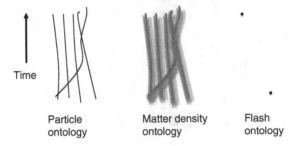

Particle   Matter density   Flash
ontology   ontology   ontology

Figure 10.5   Three Ontologies in Space-time at Microscopic Scale

tiny collection of point events (figure 10.5). These only look the same when coarse-grained at macroscopic scale.

Why would Bell suggest such a strange account of the local beables for the GRW theory given the ready availability of the matter density picture? The answer lies in his desire for a fully Lorentz invariant theory. As Tumulka eventually showed, such a theory can be developed from GRW with a flash ontology – flashy GRW for short – while few prospects exist for the theory with a matter density ontology.

The relevance of the local beables to the analysis is this: at the end of the day, what we want our theory to do is to make predictions – either deterministic or probabilistic – for the disposition of the local beables in space-time. Ultimately, it is these predictions that can be compared with the reported outcomes of experiment. In "Beables for Quantum Field Theory" Bell considers various candidates for the local beables in the context of so-called quantum "field theory," as opposed to quantum mechanics. One might naturally suppose that the local beables for a field theory ought to be something like a classical field – perhaps like our matter density – i.e. some continuous quantities defined in space. Indeed, Bell first considers the energy density, which has this field-like aspect, but rejects it for technical reasons. He settles instead on the fermion number density, with this remark:

> We fall back then on a second choice – fermion number density. The distribution of fermion number in the world certainly includes the positions of instruments, instrument pointers, ink on paper … and much much more. (Bell 1987, p. 175)

That is, the *data* that we use to test theories are determined by the distribution of fermions in space, so any theory that gets the local fermion number right will get all the data right. And exactly the same can be said of the flashes in the flash ontology. Even though the flashes are astonishingly sparse in space-time (much sparser than the fermions would be), there are enough of them to determine the positions of pointers on instruments, the configuration of ink on paper, and so on. If a theory could make correct predictions about the distribution of flashes in space-time, it would thereby make correct predictions about all macroscopic facts, and hence about all the data we use to test the theory.

What is the advantage of the flash ontology over the matter density ontology? Bell saw that even in the case of the non-relativistic theory, the flash ontology suggested the possibility of a Lorentz invariant generalization, while the matter density ontology would not. The technical term for the formal feature that the non-relativistic flashy theory has is *relative time translation invariance*, and the basic idea is as follows. Consider an experiment that has two wings that are spatially separated by a great distance. Suppose also that the two wings do not interact in the sense that the Hamiltonian of the whole system is just the sum of a Hamiltonian for one wing and a Hamiltonian for the other: there is no interaction term between the two sides. Then one can ask whether the times on the two sides "decouple": does one get the same predictions for the theory if one allows for time to advance on each side independently of the other? Put another way: in these circumstances, does absolute simultaneity, the matching of each particular moment of time in one wing with a particular moment of time in the other, play any role in the dynamics? One might expect that if the spatially distant systems really decouple, predictions made for one wing should be independent of what is happening at the other. In such a case, *knowing what happens in each wing separately should give no information about temporal relations between the wings*. Supposing one has access to all the physical facts in each wing separately – including access to all the local beables – one should not be able to determine from those facts which events in one wing are absolutely simultaneous with which events in the other. If this condition holds, then the relation of absolute simultaneity between distant events might be otiose, and therefore eliminable from the theory. Or at least, if this condition does *not* obtain, then absolute simultaneity is written into the theory at a fundamental level.

Without going into technical detail, we can ask of a non-relativistic theory: given access to the local beables of theory, could one determine which events on one wing of the experiment are absolutely simultaneous with which events on the other? If one can, then the theory does not look like a good candidate for having a Lorentz invariant generalization, and if one can't, then the existence of a Lorentz invariant version is much more plausible. What Bell showed is that the flashy version of non-relativistic GRW satisfies relative time translation invariance. But we will begin by showing that the matter density version of non-relativistic GRW does not.

Consider the following simple set-up: a single "particle," such as an electron, is put into a state that is an equal superposition of moving-to-the-right and moving-to-the-left. So long as it does not collapse, the wave-function for such a "particle" will contain two substantial parts, or lumps, which get progressively further apart as time goes on. We allow these parts to separate to a great distance and then have the electron interact, on each side, with a "position measuring device" such as the screen-plus-pointer in figure 10.4. We do not make any assumption that the event where the right-moving part of the wave-function interacts with its device is absolutely simultaneous with the event where the left-moving part interacts with its device.

The quantum-mechanical prediction for the outcome of such an experiment is simple yet profound: if we repeat such an experiment many times, about half of the time the "particle" will be "found on the right" (i.e. the pointer on the right-hand device will move to indicate a position) and about half the time the "particle" will be "found on the left." It will never be the case that the "particle" is "found" in both places, i.e. that both pointers move to indicate a location. So our experiment provides an example of an EPR correlation between the outcomes of distant experiments. Of course, such a correlation is in principle possible to explain without any spooky action-at-a-distance: simply assume that the "particle" really is a particle, with a definite location at all times, and about half the time the particle goes to the right and half the time it goes to the left. But we are here concerned with different ontologies, the matter density and flash ontologies, according to which there are no particles in this sense. In neither of these cases is there a fact about the "particle" going one way as opposed to the other before some collapse of the wave-function occurs. In such a case, as Einstein trenchantly argued, one needs some sort of spooky action-at-a-distance to assure that at most one of the two detectors will "find" the "particle".[1] Our question is whether that action-at-a-distance can be implemented in a way that respects relative time translation invariance.

Suppose we accept the matter density ontology. Then as the wave-function of the "particle" separates in the two directions, the matter density of the particle equally separates in two directions: half the density moves the right and half to the left. It is overwhelmingly unlikely that the single "particle" will suffer a GRW hit in the course of the experiment, so each of these lumps of matter density will continue to exist until one of the sides interacts with its measuring device. Suppose that in absolute time the interaction with the device on the right comes first. Then the wave-function of the device on the right will, under pure Schrödinger evolution, begin to go into a superposition of having "detected" the "particle," with the pointer moving to indicate some location on the screen, and not having "detected" the "particle," with the pointer remaining a rest in its ready state. But under the full GRW dynamics, this superposition cannot last long. By the time the

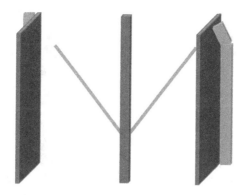

Figure 10.6   Experiment with Matter Density Ontology

two possible positions of the pointer would differ by more than 10^{-5} cm, a GRW hit on any particle in the pointer will have resolved the ambiguity, and localized the matter density of the pointer either in the initial ready position or in some position that indicates movement.

The key observation is this: since the "particle" will have become entangled with the "measuring device," the GRW hit on the pointer will *also* have the effect of localizing the previously dispersed matter density of the electron. If the hit results in the pointer having moved, then after the hit almost all of the mass density of the electron will be on the right, where the electron was "found." If the hit results in the pointer remaining in its ready position, then almost all of the matter density of the electron will suddenly appear on the left, where the "whole electron" will now appear in free flight. The electron will continue on until it interacts with the left-hand detector, which will then (with certainty) respond by moving its pointer to the location of the interaction. So in this scenario, where the right-hand interaction takes place first in absolute time, the left-hand lump of electron matter density will suffer a sudden change in mid-flight, either essentially disappearing entirely or suddenly doubling in magnitude. The former possibility is illustrated in figure 10.6.

It is evident from figure 10.6 that the GRW theory with matter density ontology is not relative time translation invariant: the synchronization of the two wings in absolute time can be read directly off the disposition of local beables in space-time. The event on the left where the matter density of the electron suddenly disappears occurs at the same moment when the interaction with the device on the right takes place. So there would be no obvious prospects for removing the explicit dependence on distant simultaneity from this theory, and hence little prospect for a relativistic theory.

The flash ontology, in contrast, yields a completely different result. Since there will almost certainly be no collapse of the electron in the course of

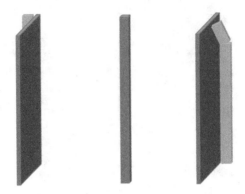

Figure 10.7   Experiment on Single Particle in Flash Ontology

the experiment, it will leave no direct traces at all in space-time. And even if it suffered a collapse, that would show up as a single point in space-time, either to the left of the source or to the right. Unlike the distribution of matter density, the single dot would give no clue about the synchronization of the two wings in absolute time.

If we imagine the flash version of figure 10.6, with enough flashes to keep track of the macroscopic objects over periods of, say, a hundredth of a second, the space-time picture looks like figure 10.7. There is no longer any indication of the relative time order of events on the right and events on the left.

What the GRW flash theory will provide is a probability measure over all possible configurations of flashes. Each of these possible configurations will either indicate that the pointer on the right has moved or that the pointer on the left has moved: none will have both pointers moving. Furthermore, the probability distribution will ascribe a 50 percent chance to configurations in which the right-hand pointer moves and a 50 percent chance to configurations in which the left-hand pointer moves. The initial state of the system (which is just a wave-function) does not determine which of these two possibilities will eventuate, and, as the EPR argument establishes, that means that the theory incorporates some non-locality even in this simple example. But the way that the indeterminism and non-locality are woven into the theory still allows for relative time translation invariance. Nothing about the resulting probability distribution of possible histories of flashes depends on which events on the right-hand side were simultaneous with which events on the left. Or, to put it another way, if we move the right-hand detector further away, so the wave-function of the electron reaches the left-hand detector first, and the collapse that determines the outcome occurs on a left-hand "particle" rather than a right-hand "particle," still the overall probabilities for the distributions of the flashes on each side will look the same, just relatively time-shifted.

Bell's final comment on the non-relativistic version of flashy GRW sums up the situation at that time:

> For myself, I see the GRW model as a very nice illustration of how quantum mechanics, to become rational, requires only a change which is very small (on some measures!). And I am particularly struck by the fact that the model is as Lorentz invariant as it could be in the nonrelativistic version. It takes away the ground of my fear that any exact formulation of quantum mechanics must conflict with fundamental Lorentz invariance. (Bell 1987, p. 209)

Despite his optimism, Bell did not formulate a fully Lorentz invariant version of the GRW collapse theory. That challenge was finally solved by Roderich Tumulka some years later. The introduction of the flash ontology by Bell, however, was one of the keys to the final solution.[2]

Relativistic Flashy GRW

Merely noting the relative time invariance of non-relativistic flashy GRW does not solve the problem of producing a fully relativistic version. The problem is that the classical space-time structure – including a universal time – is used in specifying the dynamics of the wave-function. Any fully relativistic version must somehow make due with only the resources provided by Minkowski space-time. So we should begin by reviewing exactly where the classical structure plays a role in the non-relativistic theory.

There are two obvious places where appeal is made to the classical space-time structure. One is in the probabilistic characterization of the timing for the GRW hits, or collapses. In the non-relativistic theory, each "particle" is assigned a constant probability per unit time for suffering a hit, and that probability is specified by reference to the universal classical time. Since Minkowski space-time contains no such universal time, some relativistic substitute must be found in terms of which to characterize the *when* of the collapses.

Second, the classical space-time structure is also invoked when describing the *how* of the collapses. The post-collapse wave-function is determined by multiplying the pre-collapse state by a Gaussian centered on the flash, and that Gaussian is itself defined relative to the Euclidean spatial structure of the simultaneity slices in classical space-time. Minkowski space-time also has a Euclidean structure on flat simultaneity slices, but our task is to proceed without having to utilize any such slices, which would be tantamount to picking a preferred Lorentz frame. So again, some replacement for the Euclidean hyperplanes must be found using only the resources of Minkowski space-time.

What Tumulka realized was that the Minkowski structure, together with the flashes postulated by the flash ontology, yield serviceable replacements for both universal time and Euclidean space in the non-relativistic theory (Tumulka 2006, 2009). Consider first the GRW theory for a single massive "particle." The "particle" will manifest itself in space-time as an extremely sparse collection of flashes, and the main aim of the theory is to provide, in a completely relativistic way, probabilities for where those flashes will occur. We take the initial data for the problem to be a first, or *seed*, flash and an initial wave-function for the "particle" (which may be specified relative to some arbitrarily chosen hyperplane). The theory must then specify the likelihood of the next flash being in any given region of space-time and a law for the evolution of the wave-function. Since the "particle" is massive, we postulate that all flashes subsequent to the seed flash must occur within the future light cone of the seed flash.

In the non-relativistic theory, one could arrive at the probability that the next flash occurs in some space-time region first by calculating the probability for the flash to occur at some (universal) time, and then, given that time, for it to occur within a specified spatial volume. Clearly the target time must include an interval of time – the probability for a collapse at any specified moment will be zero – and the target spatial interval must have a finite volume. Since the non-relativistic theory provides a fixed probability per unit time for a collapse, it is easy to calculate how likely the next flash is to fall within any specified interval after the initial flash. (The probability decays exponentially, like the probability for a radioactive atom to decay after some initial time.) What we need, then, is a relativistic replacement to play the role of universal time in this calculation.

But given (say) Minkowski space-time and the seed flash, it is not hard to find a serviceable analog to universal time. All points within the future light cone of the seed flash lie at some fixed invariant proper time from the flash. The surfaces of constant proper time form hyperboloids like the surface marked H in figure 4.8. These hyperboloids foliate the interior of the light cone, assigning to each event an elapsed proper time from the seed event. Since the successive flashes must lie inside the future light cone of the seed event, we may treat this foliation with its assignment of proper times exactly as we would treat universal time in the non-relativistic theory. Just as in the non-relativistic theory there is a probability for the next flash to occur at some time between t_1 and t_2, so in the relativistic theory we can define a probability for the next flash to occur somewhere between the hyperboloid at distance t_1 from the seed flash and the hyperboloid at distance t_2 from the seed flash (where the t values are the square roots of the interval between the seed flash and the next flash).

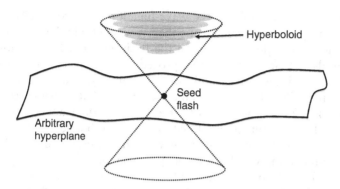

Figure 10.8 Having Chosen a Hyperboloid

Another way to put the situation is this: given the seed flash and the relativistic metric, we can define a probability density on the hyperboloids of fixed positive time-like interval from the seed. Use that probability density to choose a "time" of the next flash: it will occur somewhere on the chosen hyperboloid. This can be done using only the relativistic space-time structure. Having chosen the hyperboloid, our situation is that depicted in figure 10.8.

The data are now the position of the seed flash, the initial wave-function (expressed relative to the arbitrary hyperplane), and the chosen hyperboloid, upon which the next flash will occur. Our last problem is to give probabilities for the flash to occur in any given region of the hyperplane.

In the non-relativistic theory, one would use Schrödinger's equation to evolve the wave-function from the initial time to the time of the first collapse. In our relativistic setting, we have a relativistic equation (such as the Dirac equation) for evolving the wave-function from one hyperplane to another. In the case of a massive particle, we can evolve the initial wave-function, given relative to the arbitrary hyperplane, unitarily to a wave-function defined on the chosen hyperboloid. This can be done in a completely relativistic way requiring only the Minkowski metric.

Once we have the wave-function defined relative to the hyperboloid, we can use it to calculate probabilities for the location of the next flash on the hyperboloid in the usual way: essentially, the probability for collapse in some region is proportional to the square amplitude of the wave-function for that region. The spatial geometry of the hyperboloid is not Euclidean, even in Minkowski space-time: in Minkowski space-time, flat space-like hyperplanes have a Euclidean structure but these hyperboloids are spaces of constant negative curvature (hyperbolic geometry). But the wave-function still provides probabilities for the collapse locations in the usual way. So now we have calculated the likelihood that the next flash will occur in any given region of the future light cone of the seed.

The last task is to collapse the wave-function itself: to specify its post-collapse state. In the non-relativistic theory, we multiply the wave-function by a Gaussian function centered on the flash location. But we now have all the tools in place to adapt this procedure to a relativistic space-time. The hyperboloid has a hyperbolic spatial geometry, and we can specify a Gaussian function relative to it. Like the original GRW theory, some new constant of Nature is needed to fix the width of the Gaussian. One then multiplies the wave-function defined on the hyperboloid by the Gaussian on the hyperboloid centered at the new flash location to derive the post-flash wave-function. And the non-collapse dynamics can be used again until another flash needs to be accounted for.[3]

So far we have only discussed the dynamics for a single "particle," corresponding to a single sequence of flashes. Since each flash lies in the future or past light cone of every other, the single "particle" cannot generate any violations of Bell's inequality for experiments at space-like separation. But even so, this simplest case does shed light on our basic problem. The single "particle" does display EPR correlations at space-like separation: if a flash for this "particle" occurs in some region on a hyperboloid, there is zero chance that another flash will occur elsewhere on the same hyperboloid. If we were dealing with true particles – with continuous worldlines – this would not be puzzling: a continuous time-like curve can intersect one of our hyperboloids only once. But since the flashes are discrete, and there is no fact about "where the particle is" between flashes, there must be some non-locality in play to forbid pairs of flashes at space-like separation. As in the non-relativistic theory, this non-locality is provided by the collapse dynamics: the appearance of a flash at one location rules out the appearance of any other flash at space-like separation. So even in the one-"particle" case we have an example of non-locality implemented in a completely relativistic way.

But the big payoff comes when we expand the theory to cover a multitude of "particles." When there is no interaction Hamiltonian between the particles the extension is straightforward. The limitation to situations with no interaction Hamiltonian is rather Draconian, but still allows for situations in which Bell's inequality can be violated. For two "particles," we now begin with a two-particle wave-function (defined relative to an arbitrary hyperplane) and two seed flashes: one for each particle. One calculates the probability for the distribution of later flashes in just the same way: since there is no interaction Hamiltonian, the non-collapse evolution of the wave-function is fixed by the evolutions of the particles separately. The essential new feature is that if our pair of particles starts off in an *entangled* state, like the singlet state, then the collapse of the wave-function associated with a flash on one particle can change the probability distribution for the location of flashes of the other particle, even if it happens to be at space-like separation. So a pair of entangled "particles" in flashy relativistic GRW can exhibit behavior that violates Bell's inequality for experiments at space-like separation, even though the theory only makes use of the relativistic space-time structure in specifying its dynamics.

To model a Bell experiment, start with two electrons in the singlet state, and let them separate until there is no interaction Hamiltonian between them. (We are speaking with the vulgar as if the electrons were really particles, and have a definite location, but there is no harm: reduce this talk to a description of the usual laboratory set-up.) We use measurements of electron spin rather than the polarization of photons because we need massive particles. The statistics for correlations between spin measurements are similar to photon polarization: whenever we measure the spin of the particles in the same direction, we get opposite outcomes (one "spin up" and the other "spin down"). If the angles of the measurements are offset by an angle θ, then the outcomes disagree $\cos^2 (\theta/2)$ of the time. This leads to violations of Bell's inequality in the usual way.

The choice of a direction to measure the electron spin for each particle is represented by the free choice of one or another Hamiltonian for that particle. These Hamiltonians, local to each side of the apparatus, influence the dynamics of the wave-function and hence the probability distribution for the flashes. Just as in the case of our single particle in a superposition of going to the right and going to the left, the collapse dynamics ensures that when the spin-measuring devices are oriented in the same direction, the outcomes will always be anti-correlated: when one is up, the other will be down. This is not because the outcomes are predetermined at the source. When both measuring devices are oriented in the same direction, the initial state of the particles yields a 50 percent probability of the flashes on the right forming an "up" outcome and the flashes on the left forming a "down" outcome, and a 50 percent probability of the opposite result. There will be no chance of both devices recording "up" or both recording "down." *The same probability for the distribution of the flashes results no matter how we choose to calculate the distribution*: we can treat the right-hand interaction as occurring "first" or the left-hand interaction as "first." At the most detailed level of local beables there is – the level of flash distribution – the probabilities come out the same.

Thus, the relative time invariance of the non-relativistic theory becomes invariance under the choice of different foliations into hyperplanes in the relativistic theory. Figure 10.9 illustrates the situation: we can calculate probabilities for flashes to occur at points p (for one particle) and q (for the other) by advancing time along the A series of hyperplanes, collapsing the wave-function when the foliation passes through flash p, or by advancing along the B series, collapsing when passing through q. The probability for that particular configuration of flashes comes out the same

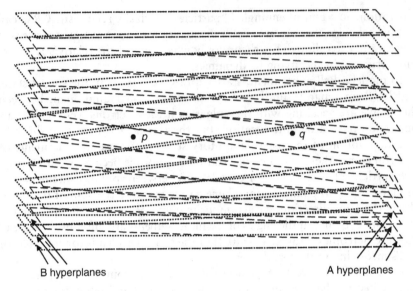

Figure 10.9   Two Foliations with Two Flashes

under either calculation, and the final wave-function with respect to the top hyperplane is also the same.

Since the probabilities come out the same no matter which foliation is used, the theory does not require that any foliation be treated as special. Only the intrinsic relativistic spatio-temporal structure is required.

When the spin-"measuring" devices are oriented at oblique angles, flashy relativistic GRW makes the same predictions for macroscopic outcomes as standard quantum theory, violating Bell's inequality. No special foliation of the space-time is employed in specifying the dynamics, and hence none could be revealed even by perfect knowledge of the complete set of flashes. Unlike the matter density ontology, the flash ontology permits a completely relativistic version of the GRW theory, and, most importantly, a completely relativistic implementation of non-locality. Relativity and quantum theory can be made compatible, although it took the combined insights of Ghirardi, Rimini, Weber, Bell, and Tumulka to work out exactly how.

The Role of Local Beables

There are several morals to the story of relativistic flashy GRW. The most obvious is that it is not sufficient to attend only to the dynamics of the wave-function. Indeed, the dynamics of the wave-function is of secondary importance in comparison with the nature and dynamics of the local beables.

The non-relativistic GRW collapse dynamics allied with the matter density ontology has no known relativistic generalization, while the flash ontology admits of a relativistic extension, as we have seen. But the ontology of local beables plays more than just an instrumental role in service of the dynamics: the local ontology is where the theory makes contact with the data. Having probabilities for wave collapses *per se* does not help in interpreting a theory until the collapse – or the quantum state itself – has some empirical significance. But wave-functions, and *a fortiori* wave-function collapses, are not directly empirically accessible. It is a general feature of quantum theory that no observation or experiment can reveal the exact state of a wave-function. So even if the wave-function were to suddenly change, it is unclear what empirical difference that could make.

When a theory has local beables, objectively existing elements in space-time, then it becomes clear how predictions about the disposition of those beables can be checked by experiment. The check does not occur at microscopic scale: as the term "microscopic" indicates, the doings at that scale lie beyond our immediate ability to observe. But physical practice is predicated on the thesis that we can observe the dispositions of macroscopic items in space and time, the directions of pointers, the outputs of experimental equipment, the trajectories of planets, and so on. And since macroscopic objects are simply large collections of microscopic entities, a theory that provides probabilistic predictions for disposition of the microscopic automatically provides probabilistic predictions for the macroscopic. It is these predictions that must be correct for the theory to be empirically successful.

Flashy GRW drives home this point with startling clarity, because the microscopic world that it posits is so far removed from anything we might have expected. We think that atoms are "just there" all the time, in some configuration or other. The astonishing output of the scanning electron microscope appears to show the locations of individual atoms – clearly enough to read "IBM" written out in individual xenon atoms on nickel. Surely, it seems, the image reveals the locations of the atoms, which are there all along.

But according to flashy GRW, there is probably *nothing at all* in space-time that spells out "IBM" at the atomic scale while the image is made. Even if we count quarks, there just aren't enough in the xenon to leave a flashy trace over the period of the experiment. Nonetheless, the macroscopic image produced by the machine will certainly come out looking as if something was there. This is guaranteed by the wave-function of the xenon and the dynamics of the theory.

Once one sees the essential role of the wave-function in carrying information about the xenon atoms, it is natural to ask what role the local beables are playing at all. If the xenon can "seem" to display a spatial structure without having any local beables, why not apply the same moral all the way through, and let the wave-function do *all* the work?

This avenue of attack runs into the problem of figure 10.3: without any local beables at all, there is no macroscopic reality in space-time. But since our information about the external world is presented to us in terms of macroscopic objects behaving in space-time, as Bohr insisted, a pure wave-function ontology leaves a theory with no empirical purchase. As little as there is to the flashy ontology on a microscopic scale, there is enough when aggregated to form recognizable macroscopic world. It is an exceptionally odd local ontology, bordering on the incredible. It comes astonishingly close to a Cartesian skeptical scenario, with the laws of physics playing the role of the evil demon: instrumentation that naively seems to provide access to the microscopic structure of the world behaves as if there are certain small localized objects even though there are none. But for all that, it is a theory that produces a familiar macroscopic world.

Flashy GRW has quite little in the way of local beables – but at the end of the day, it has enough. Other, more popular approaches, though, are taken quite seriously even though they offer no clear account of local beables at all. Most obviously, many-worlds theorists typically do not postulate any local ontology in the foundations of the theory: all there is is the wave-function. A lot of attention is paid to "observables" and "decoherence" but it is not at all clear how to generate a local ontology if all one has to work with is the wave-function. As Bell remarked, observables and observers ought to be made out of beables, out of things that exist independently of considerations about observers and "measurement." Discussions of the decoherence of the wave-function ultimately reduce to observations about the dynamics of the wave-function alone. But since the wave-function is not itself a local beable, nothing about its dynamics can yield a local ontology.

If there are no local beables in a theory – if nothing at all happens in any restricted region of space-time – then the theory cannot possibly make the same predictions as standard quantum theory, for standard quantum theory does make predictions about local matters of fact, at least at the macroscopic scale. Standard quantum theory does assert that measurements have outcomes, and provides probabilities for those outcomes. And although it is vague about exactly what constitutes a "measurement," the practice of physics recognizes many paradigm examples, such as using fluorescent screens or photomultiplier tubes to determine the "location" of an electron or a particle. At least the *outcome* of such an experiment is supposed to be a physical characteristic of the space-time region where it is performed. Standard quantum theory asserts (via the collapse postulate) that such "measurements" always *have* outcomes, and furthermore have *unique* (albeit unpredictable) outcomes. It is exactly because such experiments always have outcomes that we can ask after the predictions of the theory for the *correlations* between the outcomes: if I measure the polarization of a photon in some direction on one wing of an experiment and the polarization of an entangled photon on the other wing, how likely is it that the outcomes will be the same (both passed or both absorbed) or different (one passed and the other absorbed)? These are the phenomena that we have been trying to explain from the beginning, predicted by quantum theory and verified in the lab. But these predictions all tacitly assume that something or other perfectly definite *happens* on each wing of the experiment. If nothing at all happens, then there is nothing to be correlated, and *a fortiori* no correlation to explain.

If a many-worlds interpretation insists that there are no local beables, then this is the situation. It cannot possibly reproduce the predictions of standard quantum theory about the outcomes of experiments, and so is not relevant to our discussion of theories that agree with these predictions. But the many-worlds interpretation is never presented in this way. It is rather presented as if instead of *no* local beables, there is a (largely invisible) *profusion* of them. That is, instead of *nothing* happening on either wing of the experiment, the standard story is that *everything* happens on *both* wings: on both wings, there is "a world" in which the photon passes its polarizer and "a world" in which it is absorbed, no matter how the polarizers were oriented.

Without an account of local beables – beables that go beyond the wave-function – it is not clear how to make any sense of this claim. Insofar as there is any rough picture of what one has in mind, it perhaps seems to be something like the matter density ontology. One can imagine a theory with a matter density ontology in which the wave-function never collapses: indeed, Valia Allori, Sheldon Goldstein, Nino Zanghì, and Roderich Tumulka (2011) argue that this was essentially the theory that Schrödinger explored in 1926 (cf. also Allori *et al.* 2008). If the wave-function never collapses, then the matter density evolves into a rather indistinct blob, consisting in all the "possible" outcomes of the experiment (passed and absorbed, for example, with these results being recorded in macroscopic ways) literally superposed on one another in the same space-time region. One then tries to argue that different components of the blob are causally disconnected from one another, and so would be mutually transparent: many outcomes co-existing but unaware of each other. One will typically appeal to the decoherence of the wave-function and a functional analysis of how to separate the blob into distinct worlds to make out this conclusion.

But two facts must be kept in mind. First, as we have seen, the matter density ontology is not implied by the existence of the wave-function *per se*. Already we have seen how a wave-function with GRW dynamics can be supplemented with a matter density ontology or with a flash ontology, to very different effect. If a many-worldser wants there to be a local matter density in space-time, then that has to be postulated as something real in addition to the wave-function. Second, if we produce an account like this, then there still has to be discussion of what it means to say that outcomes on the two wings of the experiment are *correlated* to some degree. If whenever a polarization experiment is done, with any orientation of the polarizers, *both* outcomes are always produced, then it is not obvious what it might mean to say that these outcomes are correlated. If no sense can be made of this, then again the theory does not reproduce the predictions of standard quantum theory, which predicts definite correlations for outcomes at space-like separation. And if some sense can be made of the existence of correlations, we have to understand how. In particular, if appeal is made to the wave-function to explicate the sense in which, say, the "passed" outcome on the right is paired with the "absorbed" outcome on the left to form a single "world," then we have to recognize that this is not a *local* account of the correlations since the wave-function is not a local object. But these sorts of questions can be raised only once the ontology of local beables of the theory has been made explicit.

The local beables, then, provide the main subject matter of physics. The GRW wave-function dynamics provides probabilities, but as probabilities solely for the evolution of the wave-function itself they do not make immediate contact with experience or data. The probabilities that matter are probabilities for the local entities in space-time. Without them, we don't know what to make of a theory. And without them, it is hard to imagine how any theory of space-time structure, such as Relativity, could matter much. For if there isn't anything in space-time at all, it is hard to see why it could make much of a difference what the structure of space-time might be.

The Logical Situation

The discovery of relativistic flashy GRW also clarifies the relation between relativistic space-time structure and the violation of Bell's inequality. It turns out that the non-locality needed to violate the inequality can be adapted to the space-time structure: a theory written entirely in terms of relativistically kosher space-time geometry can predict violations of the inequality. This is a particularly important observation because of recent claims to a proof that this cannot be done.

In 2006, John Conway and Simon Kochen published their "Free Will Theorem" (Conway and Kochen 2006 (see also their 2009 follow-up)). This paper contains the astonishing claim of a derivation from the predictions of quantum theory to the conclusion that "if indeed there exist any experimenters with any modicum of free will, then elementary particles must have their own share of this valuable commodity" (2006, p. 1441). We will leave contentious claims about the free will of electrons aside: the strict content of the purported theorem is much simpler. Part of what Conway and Kochen prove is just this: if we treat the choices of experimenters as "free variables" – i.e., if we assume that the way the experimental apparatus on the two wings of a Bell experiment are set is uncorrelated to the initial state of particles measured – then the outcomes of each experiment cannot be a function of what happens in its back light cone. Put this way, the conclusion is already familiar: it is just Bell's conclusion that no local theory can reproduce the predictions of quantum theory for experiments at space-like separation. Bell also insisted that the experimental settings must be treated as free variables, uncorrelated to the initial state of the particles, and gave compelling physical justification for this assumption (Bell 1987, ch. 12). And the conclusion that the outcomes cannot be a function of the state on the back light cone of the experiment is just the conclusion of non-locality: if the outcomes were such a function, then the theory would be locally causal. So, at a technical level, there is nothing new in Conway and Kochen's theorem.

The particular experiment that Conway and Kochen analyze is similar in one respect to the GHZ experiment: all of the relevant probabilities needed for the proof are 1 or 0. Instead of a triple of entangled spin-1/2 particles, though, Conway and Kochen use a pair of entangled spin-1 particles. This allows them to parlay the 1967 Kochen-Specker theorem (which proved that spin measurements on a *single* spin-1 particle must be contextual) into a Bell-type non-locality proof. The basic content of this proof, however, is not anything novel.

Nonetheless, Conway and Kochen assert that their proof has some far-reaching consequences. In particular, they assert that their proof "implies that there can be no relativistically invariant mechanism of the GRW-type that explains the collapse of the wave-function" (2006, p. 1442). Assuming that all that is meant by "mechanism" here is "explicit physical theory," then we can see the problem: Conway and Kochen are asserting that they have proven the impossibility of the sort of theory that Tumulka constructed. Something has to give. If Tumulka's theory is consistent – no matter how physically implausible it might seem – then Conway and Kochen must have made a mistake.

In fact, this is just what has happened. A full examination of the flaw in the argument can be found in Tumulka (2007) and Goldstein, Tausk, Tumulka, and Zanghì (2010). But it is worth our while to investigate the care that must be taken when discussing theories that incorporate non-locality.

When they come to consider the problem of finding a relativistic version of GRW, Conway and Kochen begin their examination with these words: "Let α_0 be the information from the hits that influences the behavior of particle α" (2006, p. 1463). But already it is not at all clear how to makes sense of this in Tumulka's theory. The GRW dynamics ensures that hits, or flashes, that occur at space-like separation can be *correlated* in a way that cannot be accounted for locally, i.e. cannot be accounted for if the probabilities for the flashes are determined by the state of their past light cone, including the initial wave-function of the pair. In a local theory, the probability for a flash occurring in some region of space-time could be a function of the initial state, and of the nature of the experimental apparatus in that region, and of the sequences of flashes in the past light cone of the region. But if the theory were local, once all of these factors had been taken into account, further conditionalization on flashes *outside* the light cone could not change the probability of a flash in that region. Therefore, the joint probability of a pair of flashes in space-like separated regions would have to be the product of the probabilities of the flashes individually, conditional on the initial state and their past light cones. But we know from Bell that that no such theory can reproduce the predictions of quantum mechanics. So it is essential in the relativistic GRW theory that the probabilities for flashes in two space-like separated regions *not* always be independent: the probability of flashes occurring in both regions is not always just the product of the probability for each to occur (conditional on the state of its past light cone). The point of Tumulka's construction is to show how to generate a probability distribution over flashes that violates this independence, yet to do so in a completely relativistic way. But if we consider such a pair of regions – regions where the probability for a flash in both regions is not just the product of the probabilities for a flash in the regions taken separately – what should we say about which flash *influences* which? We would have to answer this question to specify the information α_0 mentioned above.

We cannot maintain that each flash is completely *un*influenced by the other, at least not if we draw the natural conclusion from this that their probabilities should be rendered independent once we conditionalize on their past light cones. But neither is there anything in the theory to pick out one flash in particular as influencing the other. The situation between the flashes, even at the most fundamental physical level, is perfectly symmetric. Suppose, for example, that in fact flashes occur in both of two space-like separated regions. Since the probability of the joint occurrence is not just the product of the probability of the individual occurrences (conditional on their past light cones), the occurrence of either flash *provides information* about the other: taking into account the occurrence of one flash will alter the assessment of the likelihood of the other. But should that information be considered "information from the hits that influences particle α"? All we have from the theory is a probability distribution over the set of all possible histories of flashes: the theory provides nothing more than this from which to draw conclusions about "influence." And nothing in that probability distribution implies an answer to the question of which flashes are "influenced" by which.

One way – indeed, the most obvious way – that a physical theory can be analyzed to tease out results about "influence" is to use the theory to evaluate counterfactuals. The theory, for example, might imply that the result of one experiment *would have been different* had another experiment been carried out in a different way. Since we treat the experimental arrangements as free variables, it would be natural to conclude that this would show some influence of the latter on the former. But flashy GRW never supports counterfactual conclusions of this form. Suppose, for example, we perform one of a certain class of experiments on the right and one of another class on the left. For each particular choice of experiment on the right and experiment on the left (regarded as a free choice of the experimenters), flashy GRW will provide a specific probability distribution for how the flashes throughout all space-time might be distributed. From this universal probability distribution, one can derive a probability distribution for just the flashes on the right, and a probability for just the flashes on the left. And one can further ask: does the probability distribution on one side depend at all on the choice of which experiment to perform on the other? The whole point of Bell's proof of relative time translation invariance, and of Tumulka's generalization to relativistic space-time, is to show that no such dependence exists. The probability distribution on each side is independent of what choice is made on the opposite side – even while the full probability distribution over all the flashes is *not* just the product of the distributions on each side. So the two sides exhibit information-carrying correlations, but one can reach no conclusion at all about how things would have come out on one side had a *different* experiment been carried out on the other. There is no way to apply Conway and Kochen's requirement that we consider the "information from the hits that influences the behavior of particle α." Not being able to do this, the rest of their reasoning cannot be carried through.

There is more to say about the Free Will Theorem, and a detailed analysis can be found in the papers mentioned above. But the overall point is clear: one cannot prove a contradiction between the general form of the GRW collapse theory and Relativity if a fully relativistic version of GRW detailed enough to predict violations of Bell's inequality exists. And due to the work of Tumulka, such a version does exist.

The Methodological Situation

One might have thought that the relativistic version of GRW would not only settle a certain logical question, but would also practically settle the issue of how research ought to proceed. From the beginning, our problem has been a seeming contradiction between the two great pillars of modern physics: Relativity and quantum theory. Bell proved that the tension between the theories is not just a matter of interpretation, or of misunderstanding. Insofar as one regards Relativity as implying locality – as implying, that is, that events at space-like separation must be physically independent of one another, so the probability of their joint occurrence should just be the product of the probability of their individual occurrences (conditional on everything in their back light cones) – then Relativity is indeed inconsistent with the predictions of standard quantum theory. What we have seen is that Relativity need not be interpreted in this way. If Relativity postulates only a certain space-time structure, then the question becomes whether some form of non-locality can be implemented using only that space-time structure. We now know it can be. So have we not finally discovered the Holy Grail? Should not all of our resources be directed at developing relativistic flashy GRW, or some variant thereof?

It should be noted that there would still have to be some work done on the theory to produce something capable of reproducing all the predictions of standard quantum theory. Indeed, so far relativistic flashy GRW has only been defined for *non-interacting* situations: situations, that is, where there is no interaction Hamiltonian between "particles." This is obviously a rather severe deficit: without interaction Hamiltonians, we cannot recover any of the predictions of quantum theory for atomic interactions, solid state theory, and so on. Indeed, it is not much of an exaggeration to say that the *only* interesting behavior one can derive from the theory is violation of Bell's inequality for experiments at space-like separation! That comes about because in exactly such situations, where the experiments are carried out at great distance from one another, everyone postulates that there is no interaction Hamiltonian between the sides. The lack of an interaction term is the physical manifestation of our conviction that the two sides of the experiment are practically isolated from one another, as well as the source of the conclusion that no messages can be sent from one side to the other. So the fact that relativistic flashy GRW does not yet admit interaction terms in no way affects its status as an exemplar of a fully relativistic theory capable of violating Bell's inequality. It does, however, render it almost useless for the rest of physics.

Even so, the Holy Grail is the Holy Grail: isn't this just what we have been looking for all along? Unfortunately, the situation is not so simple. I concluded the first two editions of this book with the observation that the Deity is at least extremely mischievous. We have somehow ended up in a world that seems, in its spatio-temporal aspect, to be relativistic, but also to be populated with matter governed by non-local laws. Relativity, in its usual interpretation, provides a snug and happy setting for local physics: if all of the physical facts relevant to the occurrence of an event lie in or on its past light cone, then the space-time structure itself has quarantined the relevant from the irrelevant. The speed of light would be a limit of straightforward physical significance: one would not have to look beyond the past light cone of a region to discover everything that might be predictively relevant concerning what happens there. Newton's universal time could be rejected, and the infinitesimally small regions of space-time could attend to their own business, blissfully unaware of distant events. Einstein expected a completed physics to respect this foundational locality in space-time.

Bell's result put an end to Einstein's dream. And so we have been left with a calculus of incongruities. Could it be (as some versions of the many-worlds interpretation suggest) that we can save Einstein's locality by abandoning our conviction that laboratory operations have unique outcomes? This seems like a high price to pay to retain locality. Should we give up on Relativity at a fundamental level, and postulate some preferred foliation of space-time, more congenial to Newton than to Einstein? We can, and preserve our normal opinions about macroscopic reality, but the foliation required by the theory must then also somehow be hidden from our view by the theory. We could then implement non-locality in the most obvious way, but would have to abandon the relativistic account of space-time. Relativistic flashy GRW seems to allow us to have our cake and eat it too: there is a unique local microscopic reality that yields a unique local macroscopic reality, corresponding to what we thought all along, and the whole requires only a relativistic space-time. But for all that, the incongruity pops up in a different place. For to accept this theory is to accept that microscopic reality is nothing at all like what we took it to be – not even the parts that we naively accepted as "revealed" to us by microscopes. The ball-and-stick models of DNA – which do an admirable job of accounting for cellular behavior – would be wildly misleading: a strand of DNA would show up in space-time as a sparsely scattered set of flashes, which would hardly suggest, over reasonable time periods, a double helix. Accepting the flash ontology entails rejecting the space-time picture of cellular structure that has guided the great advances in medicine and biochemistry. The theory does, of course, account for the practical success of that false picture. But the very radical falsity of the picture presents us with a methodological puzzle.

Why, in the first place, did we seek a theory that retains a relativistic account of space-time at the most fundamental level? Because that account of space-time structure has been so successful in making predictions that are checked at the macroscopic level. Space-time, in some sense, looks macroscopically relativistic, and the most satisfying explanation of this is that it is, in fact, relativistic through and through. We have always known that the empirical predictions of Relativity can be recovered by a theory that denies the fundamental truth of Relativity, a theory that postulates a preferred foliation, for example. But the empirical adequacy of such theories has never seemed to be sufficient justification. We have preferred to try to retain more than just the predictions of Relativity: we have tried to retain its spirit as well.

This desire for a fully relativistic theory is, on the face of it, puzzling. Space and time do not come under our immediate experimental gaze. Our conclusions about space-time structure are always justified by the observable behavior of matter, so one would expect any theory that gets the matter right to be perfectly acceptable. But historically, the grip of Relativity on our imagination has been stronger than this, and theories that break Lorentz invariance have been rejected out of hand even if they recover empirical Lorentz invariance at the macroscopic scale. I am not attempting to defend this judgment here: I merely note it as a matter of fact.

The problem with flashy GRW is that although it offers the possibility of retaining the relativistic account of space-time at a fundamental level, the price is the abandonment of an equally compelling picture of the disposition of local beables at the level of the cell. It is hard to shake the conviction, for example, that each individual strand of our DNA is *just there* all the time, occupying a region of space-time, splitting, recombining, and so on. This admittedly vague picture of microscopic reality is as firmly ensconced in our scientific world-picture as is any picture of space-time structure. The price flashy GRW exacts from us in order to hang on to Relativity is the abandonment of this intuitive picture of the local beables, not merely at sub-microscopic scale, where we don't think we have any access, but even at the level of collections of xenon atoms, which we think we can manipulate to spell out "IBM."

One can argue on behalf of a flashy GRW that it could in principle recover all the macroscopic data we have, including the supposed "pictures" of the xenon. Fair enough. But a theory that adds a preferred foliation to space-time can equally recover all the macroscopic results we took to be evidence for Relativity. So we are in a methodological quandary. It is hard to imagine a neutral methodological principle that could militate in favor of retaining a pre-existing theory of space-time at all costs, while allowing for the abandonment of an equally entrenched pre-existing account of the local distribution of matter at the scale which we think is probed by microscopes. The microscopic distribution of matter is not open to our direct inspection, but neither is the structure of space-time at any scale. And if you accept a many-worlds picture, then even the local *macroscopic* distribution of matter is not open to immediate inspection: most of the matter in any given region is invisible to us. So all of our options – adding a foliation, flashy GRW, many-worlds interpretation that denies unique outcomes – one way or another postulates a physical world that shields itself from our view. Each one asserts that the natural conclusions of what seem to be straightforward scientific investigations somehow go radically awry. One way or another, the world is not at all what it appears to be.

At this point in our investigation of the world, two and a half millennia after his birth, the scientific enterprise continues to confirm the insight of Heraclitus:

Φύσις κρύπτεσθαι φιλεῖ.
[Nature loves to hide herself.]

Notes

1 This sort of simple experiment with a single particle was discussed by Einstein at the 1927 Solvay conference, using a single slit to produce diffraction. Einstein pointed out that while the wave-function of the electron spread out to interact with the whole screen, the appearance of a spot on the screen in one location precludes the formation of a spot anywhere else. If the position of the spot is not determined by a pre-existent position of the electron (which would not be reflected in the wave-function) but rather involves a fundamentally stochastic process, then the fact that no second spot ever forms elsewhere requires action-at-a-distance. See Bohr 1958, pp. 41–2.

2 A very nice discussion of how different dynamics for the wave-function can be paired with different choices for the local beables of a theory can be found in Allori, Goldstein, Tumulka, and Zanghì 2008.

3 This discussion of the dynamics of the wave-function in the GRW theory is slightly simplified, but the details omitted do not change the general picture.

An Overview of Quantum Mechanics

This Overview, in the form of a set of informal notes, discusses the mathematical structure of non-relativistic quantum theory. It focuses on the fundamental mathematical structures used to describe quantum systems and to make predictions about observable properties of systems. I begin with definitions of the basic mathematical objects: linear vectors spaces, linear operators, etc., and then show how these are used in physics. Since we are not concerned with doing practical calculations most of the content which would be found in real physics texts has been left out. What remains is the structural heart of the theory, the understanding of which requires nothing more mathematically sophisticated than algebra. I have included several problems, intended to encourage the reader to play with the formalism a bit, for such play is the royal road to understanding. My aim has been to present all that non-specialists need to know about the formal structure of quantum theory.

Vector Spaces

A vector space is a collection of entities. They could be any sort of entities at all. They could be numbers, functions, arrays of numbers, arrows, graphs, pictures etc. In principle, a vector space could be made up of a collection of tables and chairs, or grains of sand. The defining conditions of a vector space make no reference at all to what the vectors are. So we simply begin with a set of entities. Vectors are represented by boldface letters.

Quantum Non-Locality and Relativity: Metaphysical Intimations of Modern Physics,
Third Edition. Tim Maudlin.
© 2011 Tim Maudlin. Published 2011 by Blackwell Publishing Ltd.

The first requirement for this set of entities to form a vector space is that there must be some addition operation defined over them, and the addition operation must be closed in the space. That is, for any two elements V_i and V_j, there must exist some unique third element V_k such that $V_i + V_j = V_k$. The integers between 0 and 10 cannot form a vector space under normal addition since it is not the case that the sum of every pair of integers in the set is itself in the set.

We should note that when we call the operation "addition" we do not require that it intuitively correspond to normal addition of numbers. The "addition" operation can be any function from pairs of elements of the set onto elements of the set. It is a black box, into which any two vectors can be put and which has certain formal features. The first feature is that whenever we put two elements in we get a unique element out. (In technical terms, addition is a total function from the $V \times V$ into V, where V is the set of vectors.)

The axioms which the addition operation must satisfy are the following:

For arbitrary V_i, V_j, and V_k of V
(i) $V_i + V_j = V_j + V_i$ (commutativity)
(ii) $V_i + (V_j + V_k) = (V_i + V_j) + V_k$ (associativity)
(iii)   There exists a unique vector 0 such that $0 + V_i = V_i + 0 = V_i$
(iv)   For each V_i there is a unique inverse $(-V_i)$ such that $V_i + (-V_i) = 0$

The set of positive integers cannot form a vector space under normal addition since it does not satisfy Axioms iii or iv. The set of non-negative integers does not satisfy iv.

Beside being able to add vectors to one another, you must also be able to do one more thing: multiply them by a number (scalar). In this case, three axioms must obtain. Scalars are represented by Greek letters.

For arbitrary V_i, V_j, α and β,
(v) $\alpha(V_i + V_j) = \alpha V_i + \alpha V_j$
(vi) $(\alpha + \beta)V_i = \alpha V_i + \beta V_i$
(vii) $\alpha(\beta V_i) = (\alpha\beta)V_i$

Any set of objects which satisfies axioms i through vii is a *linear vector space*. That's all there is to it.

When we said that you must be able to multiply a vector by a number (scalar), we did not specify just what kind of numbers were involved. There are two main choices (although other choices are possible): the *real* numbers and the *complex* numbers. (The complex numbers are the numbers which can be represented by $a + bi$, where a and b are real numbers and i is $\sqrt{-1}$.) The domain of the allowed scalars is called the *field* of the vector space, so there are real vectors spaces and complex vectors spaces. In quantum mechanics we use complex vector spaces.

The set of integers cannot form a real vector space even though it satisfies all of the axioms for addition since an integer multiplied by a real number need not be an integer.

The definition of a vector space is pretty minimal. If you want to have a set of objects which can be added together and multiplied by scalars, the seven axioms above would intuitively all have to be met. Note that there is nothing in the definition of a vector space which requires that a multiplication operation be defined over the vectors.

Even though the definition of a vector space is very weak and abstract, there are many things which can be proven from the axioms. The first exercise is to prove the following:

Problem 1:   Prove that $-0 = 0$, i.e. that 0 is its own inverse

Problem 2:   Prove that $0V_i = 0$ for all V_i (hint: add $0V_i$ to αV_i)

Problem 3:   Prove that $\alpha 0 = 0$ (hint: add $\alpha 0$ to αV_i)

Problem 4:   Prove that $(-1)V_i = (-V_i)$ (hint: add V_i to $(-1)V_i$).

Some Examples

Vector spaces are defined very abstractly, but very simply. In order not to be misled by the properties of some single representation of a vector space, we will consider several quite different examples.

Arrows

When in high school, one learned that vectors are represented by arrows, objects with a magnitude and direction that live in some Euclidean space. The notion of a vector is actually defined much more generally, although arrows in a Euclidean space do provide an example of a vector space.

Suppose we start with the set of arrows of all possible directions and magnitude that can be drawn in a Euclidean space of some dimension. The addition operation for the arrows is the familiar head-to-tail procedure: put the tail of the second arrow to the head of the first and the resulting sum is the arrow which goes from the tail of the first to the head of the second (figure 1). It is easy to verify that this addition function for arrows satisfies the first four axioms, with an arrow of zero length being the identity and the inverse of an arrow being an arrow of equal length oppositely oriented.

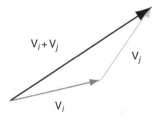

Figure 1 Addition of Arrows

It is easiest to create a vector space for arrows whose field is real numbers. All we need to know is what happens when we multiply an arrow by a real number. Such multiplication simply changes the length of an arrow, leaving its direction unchanged. Thus, multiplying an arrow by 0.5 yields a parallel arrow half as long, multiplying by –2 gives an arrow twice as long but oppositely oriented. This definition of multiplication by a scalar obviously satisfies axioms v–vii.

(Arrows can also be used to define a complex vector space, but a geometrical interpretation of this space is not so obvious. The problem comes in interpreting the result when one multiplies an arrow by $\sqrt{-1}$. To solve this problem the arrows must have a third property beside length and direction. This third property is called the phase.)

Functions

Consider the set of all continuous real functions of one variable, x. The functions form a vector space under the addition defined by $(F_i + F_j)(x) = F_i(x) + F_j(x)$. The identity element is the constant function $F(x) = 0$. It is easiest to visualize the addition graphically as in figure 2. Multiplication by a scalar is also obvious. If the functions are real valued then we use the real numbers as our scalar field and $(\alpha F)(x) = \alpha(F(x))$.

(If we want a complex vector space the natural thing to do is use complex valued functions of x, that is, functions whose values are complex numbers. Furthermore, we may want to restrict the set of functions allowed as vectors. In particular, consider the set of complex-valued functions of x such that $\int_{-\infty}^{\infty} |F(x)|^2 \, dx$ is finite. The sum of any two such functions is another, and the whole class forms a complex vector space. This vector space is called *Hilbert space*, and is the natural structure for formulating quantum mechanics. We may also use functions of more than one variable.)

Matrices

A matrix is a rectangular array of numbers. If we have two such arrays of the same dimensions we can add them by adding the corresponding

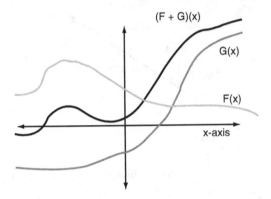

Figure 2 Adding Functions

entry in each array. For example, $\begin{bmatrix} 2 & .4 \\ 0 & -2 \\ .5 & 1 \end{bmatrix} + \begin{bmatrix} -2 & 3 \\ -1 & 0 \\ .8 & .2 \end{bmatrix} = \begin{bmatrix} 0 & 3.4 \\ -1 & -2 \\ 1.3 & 1.2 \end{bmatrix}$.

Multiplication by a scalar is achieved by multiplying each entry by the scalar: $2 \times \begin{bmatrix} 2 & .4 \\ 0 & -2 \\ .5 & 1 \end{bmatrix} = \begin{bmatrix} 4 & .8 \\ 0 & -4 \\ 1 & 2 \end{bmatrix}$. The collection of all matrices of a given size whose

entries are real numbers forms a real vector space. The collection of all matrices of a given dimension whose entries are complex numbers forms a complex vector space. The identity element is the matrix all of whose entries are 0.

Arrows, functions, and matrices do not seem very similar at first glance, but all can be vectors given the appropriate definition of vector addition and of multiplication by a scalar. In fact, quantum mechanics was originally developed in two different ways, one way using matrices ("matrix mechanics") and the other using functions ("wave mechanics"). Only later did Dirac demonstrate that the two formulations were just different ways of writing the same theory. They both employed the same vector space, one using matrices as the vectors, the other using functions. In quantum mechanics, the physical state of a system is represented by a vector in a vector space.

Linear Operators

A linear operator is a sort of mapping from a vector space onto itself. That is, it is a function which takes a vector as input and gives a vector as output. In order to be linear, an operator Ω must satisfy the following constraints:

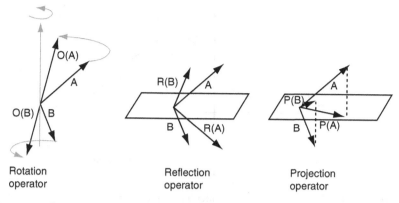

Rotation operator	Reflection operator	Projection operator

Figure 3   Three Linear Operators

For all V_i and V_j in V,
$$\Omega(V_i) + \Omega(V_j) = \Omega(V_i + V_j)$$
$$\Omega(\alpha V_i) = \alpha(\Omega V_i).$$

If an operator is linear it doesn't matter whether we first apply the operator to two vectors and then add, or first add the vectors and then apply the operator; the result will be the same.

Examples of linear operators
In the vector space of arrows, many of the linear operators can be easily visualized. For example, consider the vector space of arrows in three dimensions. Choose an axis. Rotation of the vector by a fixed amount around that axis is a linear operator. Or take the same space and choose a plane. Mirror reflection of arrows through that plane is a linear operator. Finally, choose a plane and imagine the shadow that an arrow casts on that plane if light is shining down perpendicular to the plane. The function which maps each arrow onto its shadow is a linear operator, called the projection onto the plane (see figure 3).

There are two linear operators on functions which are of particular importance for quantum mechanics. One is the function which maps F(x) to xF(x), so that $\Omega(\sin(x)) = x \sin (x), \Omega(x) = x^2$, and so on. The other is the differentiation function d/dx which gives the slope of the function. These operators are illustrated in figure 4.

Problem 5:   Show that the squaring operator Sq(F(x)) = F(x)^2 is not a linear operator. (Hint: it suffices to show that the definition fails for two particular functions.)

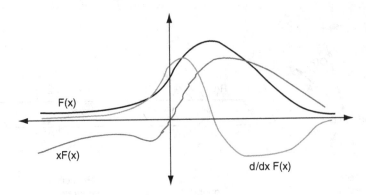

Figure 4 A Function and its Image Under Two Operators

For matrices, operators are generally represented by other matrices using matrix multiplication. A matrix of M rows and N columns can be multiplied with a matrix of N rows and P columns to yield a matrix of M rows and P columns. So, for example, a 2 × 2 matrix multiplied with a 2 × 1 yields another 2 × 1 matrix. A 2 × 2 matrix is therefore an operator on the space of 2 × 1 matrices.

The rule for multiplying matrices is as follows. To calculate the entry on the ith row and the jth column of the product matrix, go *across* the ith row of the first matrix and *down* the jth column of the second, multiplying the corresponding entries, and then sum up all of those products. Since we will mostly be concerned with 2 × 1 and 2 × 2 matrices, we will show the general cases:

$$\begin{bmatrix} a & b \\ c & d \end{bmatrix} \times \begin{bmatrix} e \\ f \end{bmatrix} = \begin{bmatrix} ae + bf \\ ce + df \end{bmatrix}$$

$$\begin{bmatrix} a & b \\ c & d \end{bmatrix} \times \begin{bmatrix} e & f \\ g & h \end{bmatrix} = \begin{bmatrix} ae + bg & af + bh \\ ce + dg & cf + dh \end{bmatrix}$$

We will see several important examples of operators on matrices anon. In anticipation, consider a vector space of 1 × 2 matrices. The matrix $\begin{bmatrix} 1 & 0 \\ 0 & -1 \end{bmatrix}$ is a linear operator on this space, mapping the vector $\begin{bmatrix} a \\ b \end{bmatrix}$ to the vector $\begin{bmatrix} a \\ -b \end{bmatrix}$. The matrix $\begin{bmatrix} 0 & 1 \\ 1 & 0 \end{bmatrix}$ is another linear operator, mapping $\begin{bmatrix} a \\ b \end{bmatrix}$ to $\begin{bmatrix} b \\ a \end{bmatrix}$.

The Eigenvalue Problem

Suppose we are given a vector space V and an operator Ω. A useful question to ask, as it turns out, is whether there are any vectors in V such that when the operator operates on the vector the result is the vector itself multiplied by some scalar. That is, are there any vectors V and scalars N such that $\Omega V = NV$. This problem is called the eigenvalue problem. Every vector which satisfies the equation is called an *eigenvector* of the operator, and the value of the scalar N for which the equation is satisfied is called the *eigenvalue* of the eigenvector.

Let's consider the eigenvalue problem for some of the operators given above. First, let's take rotation of 45° around the z-axis in the vector space of arrows. In order to satisfy the eigenvalue equation, an arrow would have to have its direction unchanged by the rotation. The only arrows for which this is true are those which point in the direction of the z-axis itself. All of those arrows are eigenvectors of the rotation. The eigenvalue for all of the vectors is 1 since they are completely unchanged by the rotation.

In the case of the reflection operator, there are two classes of eigenvectors. First, arrows which lie in the plane of reflection are unchanged by the reflection, and so are eigenvectors with eigenvalue 1. Any arrow which is perpendicular to the plane of reflection is exactly reversed by the operator. These vectors are eigenvectors with eigenvalue –1.

The projection operator is rather like the reflection operator. Again, any vector lying in the plane of projection is unchanged by it, and so is an eigenvector with eigenvalue 1. Perpendicular arrows are mapped into the vector 0. They therefore also satisfy the eigenvalue equation with eigenvalue 0.

We are now in a position to understand the basic mathematical structure of quantum mechanics. As we have already said, in quantum theory the physical state of a system is represented by a vector in a vector space. How do these vectors carry information about the behavior of the system? The connection to observable behavior is made by a second postulate: *all observable quantities can be represented by linear operators on the vector space.* In the formulation which uses wave functions, for example, the observable *position* is associated with the multiplication by x, and the observable *momentum* with differentiation.

Several further postulates can now be stated. The most important is that *when an observable quantity is measured, the only possible values which can be found for that quantity are the eigenvalues of the associated operator.* It is from this postulate that the quantization of observable quantities may be demonstrated, as we will see. A second postulate is that *a system is certain to give a particular measured value of an observable if and only if its state is an eigenvector of the observable with that eigenvalue.*

To illustrate these postulates we will present the mathematical formulation of the theory of spin. The spin of a particle is an intrinsic magnetic moment associated with the particle. The vectors used to represent the spin state of spin–½ particles are the 1×2 complex matrices, that is, 1×2 matrices both of whose elements are complex numbers. These spin vectors, or *spinors*, may therefore be generally represented as $\begin{bmatrix} a \\ b \end{bmatrix}$ where a and b are complex numbers, or $\begin{bmatrix} a+bi \\ c+di \end{bmatrix}$, where a, b, c, and d are real numbers.

What about the observables? The spin of a particle in any direction can be observed by passing the particle through an inhomogeneous magnetic field oriented in that direction. The possible results of such an experiment are the eigenvalues of the operator which represents spin in that direction. So we only need to know the operators for spin in a given direction. As we have seen, the linear operators on this vector space are 2×2 matrices.

The operator for spin in the z-direction is $\begin{bmatrix} 1 & 0 \\ 0 & -1 \end{bmatrix}$. The operator for spin in the x-direction is $\begin{bmatrix} 0 & 1 \\ 1 & 0 \end{bmatrix}$, and for spin in the y-direction $\begin{bmatrix} 0 & i \\ -i & 0 \end{bmatrix}$. Spin in the various intermediate directions is represented by other matrices.

If we carry out a spin measurement in the z-direction, the only possible results are the eigenvalues of $\begin{bmatrix} 1 & 0 \\ 0 & -1 \end{bmatrix}$. So we need to know for what values of N can the equation $\begin{bmatrix} 1 & 0 \\ 0 & -1 \end{bmatrix} \times \begin{bmatrix} a \\ b \end{bmatrix} = N \begin{bmatrix} a \\ b \end{bmatrix}$ hold. Carrying out the multiplication we get the equation $\begin{bmatrix} a \\ -b \end{bmatrix} = N \begin{bmatrix} a \\ b \end{bmatrix}$, which is equivalent to the pair of equations a = Na and –b = Nb. For what values of N can these equations be simultaneously solved?

Clearly the first equation can hold only if N = 1, assuming that a is not zero. If N = 1, then the second equation can only hold if b = 0. So 1 is an eigenvalue of the operator, with the eigenvectors being $\begin{bmatrix} a \\ 0 \end{bmatrix}$, for any value of a. If b is not zero, then the second equation can only hold if N = –1. In this case a must be 0 for the first equation to hold.

In sum, the eigenvalues of $\begin{bmatrix} 1 & 0 \\ 0 & -1 \end{bmatrix}$ are 1 and −1, with the corresponding

eigenvectors being $\begin{bmatrix} a \\ 0 \end{bmatrix}$ and $\begin{bmatrix} 0 \\ b \end{bmatrix}$ respectively. (We commonly use normalized

vectors to represent the state of systems, that is, vectors such that $|a|^2 +$

$|b|^2 = 1$, so the eigenvectors are $\begin{bmatrix} 1 \\ 0 \end{bmatrix}$ and $\begin{bmatrix} 0 \\ 1 \end{bmatrix}$.) If we measure the spin of a par-

ticle in the z-direction we can only get two possible results, 1 and −1. Thus spin is quantized; it cannot, as classical spin does, take on a continuous range of values. The only systems which are certain to yeild the z-spin value

1 are those in the spin state $\begin{bmatrix} 1 \\ 0 \end{bmatrix}$, and only those in state $\begin{bmatrix} 0 \\ 1 \end{bmatrix}$ are certain to

give value −1. Systems in any other state are in states of indefinite z-spin. If their spin is measured in the z-direction they may give value 1 or value −1. Calculating the probabilities for these different possibilities requires a bit more machinery than we have introduced so far.

Problem 6: What are the eigenvalues of the x-spin operator $\begin{bmatrix} 0 & 1 \\ 1 & 0 \end{bmatrix}$? What

are the eigenvectors associated with these values? What are the eigenvalues

of the y-spin operator $\begin{bmatrix} 0 & i \\ -i & 0 \end{bmatrix}$? What are the eigenvectors associated with

these values?

Complex Numbers

In the first section of this overview we found that every vector space has a field, i.e. a class of scalars by which the vectors can be multiplied. The usual fields are the real numbers and the complex numbers, yielding real and complex vector spaces respectively. We also mentioned that in quantum theory the physical states of systems are represented by vectors in a complex vector space. Since the role of the complex numbers becomes more pronounced as we proceed, we will begin with a short review of complex numbers.

A complex number is commonly represented as a + bi, where a and b are real numbers and i is $\sqrt{-1}$. This representation allows us to add and multiply the numbers: (a + bi) + (c + di) = (a + b) + (c + d)i, and (a + bi) × (c + di) = (ac − bd) + (bc + ad)i. There is, though, a somewhat more intuitive way to represent and visualize complex numbers and the arithmetic operations on them.

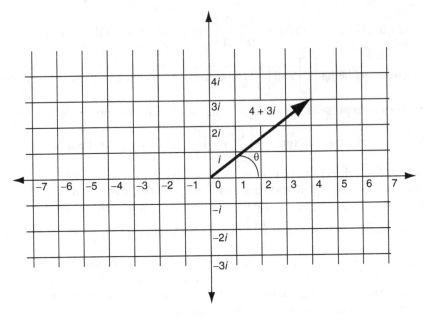

Figure 5   Complex Number Displayed on a Plane

We begin by setting up a Cartesian coordinate system with one of the axes representing the real part and the other representing the imaginary part of the complex number. In figure 5, the complex number $4 + 3i$ is shown as a vector on this Cartesian grid. It is obvious that instead of characterizing the complex number as having a real and an imaginary part we could instead convey the same information by specifying the length of the vector and the angle θ that it makes with the real axis. The vector shown, for instance, has length 5 and makes an angle $\tan^{-1}(3/4)$ with the real axis. In general, then, a complex number can be represented by the ordered pair $< L, \theta >$ where L is the length and θ the angle. L can be called the *magnitude* of the complex number and θ is a second quality called the *phase*.

What advantages accrue to this new representation? Addition of complex numbers becomes a bit more difficult: the sum $< L, \theta > + < L', \theta' >$ is not a simple function of L, L', θ, and θ'. But multiplication of complex numbers in this representation becomes simple and intuitive: $< L, \theta > \times < L', \theta' > = < LL', \theta + \theta' >$. That is, the magnitude of the product is just the product of the magnitudes and the phase of the product is the sum of the phases. Figure 6 shows the result of a multiplication graphically.

Problem 7:   Prove that the magnitude of the product of two complex numbers is the product of their magnitudes. (Hint: go back to the definitions

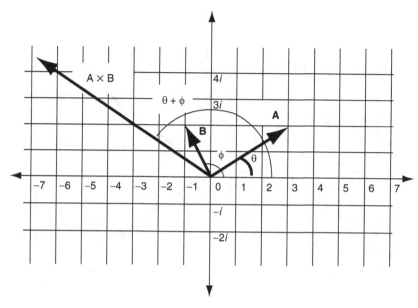

Figure 6 A × B

in terms of a + bi.) Optional for the ambitious: show that the phase of the product is the sum of the phases.

Instead of being written as < L, θ >, complex numbers are written as $Le^{i\theta}$. This is arithmetically accurate since $e^{i\theta} = \cos θ + i \sin θ$, recovering the correct expression in the form a + bi. (To prove this last equality, expand out $e^{i\theta}$ as a power series and compare it to the power series expansions of sine and cosine.) In this representation the structure of multiplication is obvious: $Le^{i\theta} \times L'e^{i\theta'} = LL'e^{i\theta}e^{i\theta'} = LL'e^{i(\theta + \theta')}$.

Every complex number C has a *conjugate* represented by C*. If we write C as a + bi, the conjugate is a − bi. If we write C as $Le^{i\theta}$, C* is $Le^{-i\theta}$. The equivalence of these two definitions is obvious from figure 7: flipping the imaginary part to the negative of its value is the same as rotating the vector by −θ rather than θ. The operation of conjugation allows us easily to define the magnitude of a complex number: the magnitude is $\sqrt{CC^*}$.

Problem 8: Prove that the magnitude of the complex number $Le^{i\theta}$ is $\sqrt{Le^{i\theta}\left(Le^{i\theta}\right)^*}$ using the magnitude/phase representation. (Hint: use the fact that $e^0 = 1$.) Prove that the magnitude of the complex number a + bi is $\sqrt{(a+bi)(a+bi)^*}$ using that representation.

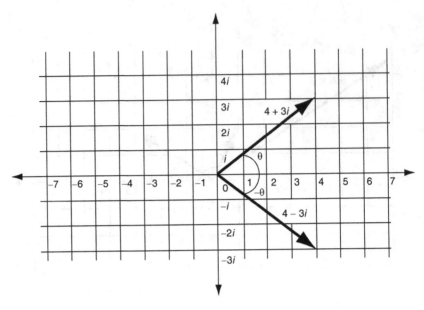

Figure 7   Conjugate Complex Numbers

With these facts about complex numbers in hand, let us now return to vector spaces.

Dimensionality of a Vector Space

Suppose you are given a (finite or infinite) set of vectors $\{V_1, V_2, V_3, \ldots\}$ and another vector W. W is said to be *linearly independent* of $\{V_1, V_2, V_3, \ldots\}$ if there are no scalars $\alpha_1, \alpha_2, \alpha_3, \ldots$ such that $W = \alpha_1 V_1 + \alpha_2 V_2 + \alpha_3 V_3 + \ldots$. A set of vectors is said to be a linearly independent set if each vector is linearly independent of the remainder of the set. A necessary and sufficient condition for a set of vectors $\{V_1, V_2, V_3, \ldots\}$ to be linearly independent is that the only solution to the equation $\sum_i \alpha_i V_i = 0$ occurs when each of the α_i is 0.

Problem 9:   Prove that the condition just given is a necessary and sufficient condition for each member of the set to be linearly independent of the rest. (Hint: postulate in turn that each side of the biconditional is false and prove in each case that the other side is.)

Definition: A vector space is *n-dimensional* if it admits at most *n* vectors that are linearly independent.

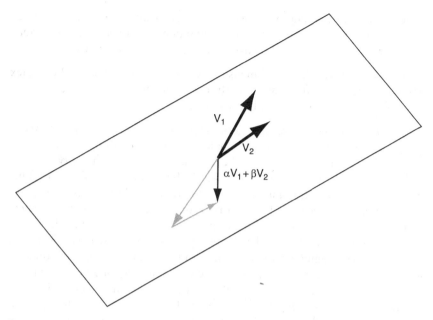

Figure 8   Two Vectors and Those Linearly Dependent on Them

Let's check that this definition of dimensionality fits the cases for which we already have intuitions. Take the vector space of arrows in three-dimensional Euclidean space. Choose a vector V_1 at random. The only vectors which are *not* linearly independent of V_1 are those which can be expressed as αV_1, i.e. those vectors which point in the same direction as V_1. So let's now choose a second vector V_2 which is not parallel to V_1. V_1 and V_2 are linearly independent. Which vectors are *not* linearly independent of the pair $\{V_1, V_2\}$? Any vector which can be written as $\alpha_1 V_1 + \alpha_2 V_2$ is not linearly independent of them. It should be obvious upon reflection that the set of vectors which are not linearly independent of V_1 and V_2 are the vectors which lie in the plane defined by V_1 and V_2. An example is shown in figure 8. The set of vectors which are linearly dependent on V_1 and V_2 form a vector space. It is easy to check that all of the defining conditions for a vector space hold: it is closed under addition of vectors and multiplication by scalars, includes the 0 vector, includes inverses of all its members, etc. In technical terms, vectors in the plane form the *subspace* which is *spanned* by the set $\{V_1, V_2\}$.

Still, the set $\{V_1, V_2\}$ is not the largest set of linearly independent vectors which can be found in the three-dimensional vector space. Obviously, any vector which does not lie in the plane is linearly independent of $\{V_1, V_2\}$. Choose one of these vectors V_3. It should again be obvious that any vector at all in the space can be expressed as $\alpha_1 V_1 + \alpha_2 V_2 + \alpha_3 V_3$. (The proof of this obvious fact is rather lengthy, so we omit it.) Since any vector can be written as $\alpha_1 V_1 + \alpha_2 V_2 + \alpha_3 V_3$, no further linearly independent vector can be added to the set $\{V_1, V_2, V_3\}$. The dimension of this space is therefore 3.

Problem 10:   Prove that the dimension of the vector space of 1×2 complex matrices is 2. (Hint: find two linearly independent vectors in the space in terms of which any vector can be expressed.)

The dimensionality of the spaces of arrows and of matrices are intuitively pretty obvious. What is the dimensionality of the space of functions? That is, how large a set of functions $\{f_1, f_2, f_3, \ldots\}$ must we have so that *any* function could be expressed as $\alpha_1 f_1 + \alpha_2 f_2 + \alpha_3 f_3 + \ldots$? The answer is that the set of functions must be infinite. It is a little difficult to prove this, but here is an intuitive sketch. In order to specify a 1×2 matrix I must give you two numbers. There are two independent degrees of freedom in the matrix, and hence the vector space is of dimension 2. In order to specify an arrow in Euclidean 3-space I have to give you 3 numbers; there are three independent degrees of freedom. In order to specify a function of x, I must give you an infinity of numbers, namely the value of the function for all possible values of x. A function therefore has an infinite number of independent degrees of freedom and the vector space of functions is infinite-dimensional. Fourier analysis shows that any function can be expressed as the sum of sine and cosine functions of various periodicities. Since there are an infinite number of such sines and cosines, all linearly independent of one another, the space of functions must be infinite-dimensional. Hilbert space is therefore an infinite-dimensional vector space.

The Inner Product

Consider once again the vector space of arrows in Euclidean space that we all know and love. So far we have captured in formal terms several aspects of that space: we have defined how arrows add and what happens when they are multiplied by scalars. But there are still several familiar features which have not been explicated. In particular, we have said nothing about what determines the *length* of an arrow or the *angle* between two arrows. In general, we have not defined magnitudes or angles between vectors in any of our vector spaces. It is easy to overlook this fact in the case of the arrows since we are so used to picturing their lengths and angles automatically, but perhaps the problem is clearer with functions and matrices. What is the magnitude of the vector $1/x^5$? What is the angle between the vector $1/x^5$ and the vector $\sin(x + 4)$? What is the angle between the vector $\begin{bmatrix} 3 \\ i \end{bmatrix}$ and the vector $\begin{bmatrix} 2 - 3i \\ 4 \end{bmatrix}$?

In fact, if all we have done is define the addition function and multiplication by a scalar, we have not yet provided enough structure to define magnitudes and angles. The extra piece of machinery we need is called the *inner product*. The inner product is a mapping from pairs of vectors onto scalars (from $V \times V$ onto the field F). We will represent the inner product using angled brackets and dropping the boldface for the vectors. Thus the inner product of $\mathbf{V}_i$ and $\mathbf{V}_j$ is $< V_i \mid V_j >$ (this anticipates the Dirac bra-ket notation used in quantum mechanics).

The inner product must satisfy the following three axioms:

(i) $< V_i \mid V_i > \geq 0$ (0 only if $\mathbf{V}_i = 0$)
(ii) $< V_i \mid V_j > = < V_j \mid V_i >^*$
(iii) $< V_i \mid \alpha V_j + \beta V_k > = \alpha < V_i \mid V_j > + \beta < V_i \mid V_k >$

Notice that the inner product is *not* commutative (by ii), so we must be careful to specify which vector goes into the first slot and which goes into the second. The Dirac bra-ket notation always keeps this clear: the vector in the bra $< V_i \mid$ goes into the first slot and the vector in the ket $\mid V_j >$ into the second. (To be technically more precise, the bra is a co-vector and the ket a vector, but let's not worry about that.)

Condition iii says that the inner product is linear in the second vector. It is not linear in the first vector, but rather *antilinear*, that is,

(iv) $< \alpha V_i + \beta V_j \mid V_k > = \alpha^* < V_i \mid V_k > + \beta^* < V_j \mid V_k >$

When we multiply the second vector by a scalar the inner product just gets multiplied by the scalar, but when we multiply the first vector by a scalar the inner product is multiplied by the complex conjugate of the scalar. If the vector space is a real vector space then the inner product is commutative (since a real number equals its complex conjugate) and we don't have to worry about the order of the vectors any more.

Condition iv is not an independent condition from the other three. In fact, it can be proven from the other three as follows:

$$< \alpha V_i + \beta V_j \mid V_k > = < V_k \mid \alpha V_i + \beta V_j >^* \qquad \text{by (ii)}$$

$$= \left(\alpha < V_k \mid V_i > + \beta < V_k \mid V_j > \right)^* \qquad \text{by (iii)}$$

$$= \alpha^* < V_k \mid V_i >^* + \beta^* < V_k \mid V_j >^* \qquad \text{by complex arithmetic}$$

$$= \alpha^* < V_i \mid V_k > + \beta^* < V_j \mid V_k > \qquad \text{by (ii)}$$

Once we have provided our vector space with an inner product it becomes an *inner product space*.

We can now define the magnitude of a vector, which is called its *norm* $|V|$:

$$|V| = <V|V>^{1/2}$$

A vector is said to be *normalized* and is called a *unit vector* if its norm is 1.

We can also use the inner product to define an angle between vectors. In practice, we often care about one question: are the vectors perpendicular (at right angles) to one another or not. Two vectors V_i and V_j are perpendicular if $<V_i|V_j> = 0$. Such vectors are also said to be *orthogonal* to one another.

What are the inner products usually used in our three vector spaces?

First consider the space of arrows in Euclidean 3-space. In Euclidean space there is an angle θ between any two (non-zero) vectors, and the inner product $<V_i|V_j> =$ (length of V_i) $\times$ (length of V_j) $\times \cos(\theta_{ij})$. Another way of picturing the inner product here is as follows. Introduce a Cartesian coordinate system into the Euclidean space. Every arrow now becomes associated with a triple of real numbers $<x, y, z>$. The inner product $<V_i|V_j> = x_ix_j + y_iy_j + z_iz_j$.

In Hilbert space, the inner product of two vectors $f(x)$ and $g(x)$ is $<f(x)|g(x)> = \int_{-\infty}^{\infty} f(x)^* g(x)dx$. This explains why Hilbert space contains only the square-integrable functions: these are just the functions with finite norms.

In the space of 1×2 complex matrices, the inner product is calculated as follows. First, take the complex conjugate of the transpose of the first vector. (The transpose of a 1×2 matrix is the 2×1 matrix whose first entry is the top entry of the given matrix and whose second entry is the bottom entry of the given matrix, i.e. $\begin{bmatrix} a \\ b \end{bmatrix}^T = [ab]$.) Then multiply this by the second matrix. It is easier to see than to describe:

$$<\begin{bmatrix} a \\ b \end{bmatrix}|\begin{bmatrix} c \\ d \end{bmatrix}> = \begin{bmatrix} a^*b^* \end{bmatrix} \times \begin{bmatrix} c \\ d \end{bmatrix} = a^*c + b^*d.$$

Problem 11:   Calculate the inner product of the eigenvectors for z-spin with the eigenvectors for x-spin. Calculate the inner product of the eigenvectors for y-spin with the eigenvectors for x-spin.

Orthonormal Bases

We now have all of the machinery we need to calculate probabilities in quantum mechanics. But before going on to do this, we can use a few more definitions.

A set of linearly independent vectors which span a vector space is called a *basis* for that space. Any vector space will have an infinite number of different bases. If, further, for all V_i and V_j in the basis $< V_i|V_j > = \delta_{ij}$ then the basis is said to be *orthonormal* (where the *Kronecker delta symbol* δ_{ij} is 1 if $i = j$ and is 0 if $i \neq j$.) In normal English, a basis is orthonormal if the norm of each basis vector is 1 and if each basis vector is orthogonal to every other one.

Now that we have the inner product we can also pick out a special class of linear operators. Suppose we have an operator Ω. In general, there is no reason to expect that the inner product of V_i with ΩV_j would equal the inner product of ΩV_i with V_j. If it is always true that $< \Omega V_i|V_j > = < V_i|\Omega V_j >$ then Ω is said to be *Hermitian*.

Problem 12: Prove that the eigenvalues of a Hermitian operator are all real (Hint: consider the inner product $< V_i|\Omega V_i >$ where V_i is an eigenvector of Ω. Use the eigenvalue equation $\Omega V_i = nV_i$ and the fact that $n = n^*$ only if n is real.)

We can now be a bit more precise than we were originally about observables in quantum mechanics. We said that observables are all associated with linear operators and that the possible results of an observation are the eigenvalues of the operator. But in principle there is no reason why the eigenvalue of an operator might not be a complex number, and one might be puzzled as to how the result of an observation of, say, position or momentum might be a complex number. In fact, observable quantities in quantum mechanics are associated with Hermitian operators, and hence with operators all of whose eigenvalues are real. Furthermore, it is generally assumed (although this is rather controversial) that every Hermitian operator corresponds to a quantity which can in principle be observed.

We need just one more fact, which I will cite without proof. *The eigenvectors of a Hermitian operator span the vector space.* In fact, in most cases there is a unique (up to a phase) orthonormal basis of the vector space all of whose members are eigenvectors of a given operator. (The complication arises when there are linearly independent eigenvectors with the same eigenvalue, in which case the operator is said to be degenerate. In this case there will be an orthonormal basis of eigenvectors, but it will not be unique.)

We can now state the main postulate of quantum theory. Suppose a quantum system is in the state represented by a normalized vector V and you are going to carry out on the system a measurement represented by the Hermitian operator Ω. What does quantum mechanics predict?

We already know that the only possible results of the measurement are the eigenvalues of Ω. If the operator is not degenerate, every eigenvalue n is associated with a unique normalized eigenvector N. *The probability that the measurement will yield the value n is just* $| < N|V > |^2$, that is, it is the absolute square of the inner product of V and N. (The absolute square of a number C is CC^*.)

Here is another way of looking at the situation. Associated with every non-degenerate Hermitian operator Ω is a set of eigenvectors $N_1, N_2, N_3, \ldots$ which form an orthonormal basis of the vector space. This means that any given vector V can be expressed as $\alpha_1 N_1 + \alpha_2 N_2 + \alpha_3 N_3 + \ldots$. The probability of getting result n_i if we perform the measurement associated with Ω on a system in state V is just $|\alpha_i|^2$.

Problem 13: The normalized eigenvectors of z-spin are $\begin{bmatrix} 1 \\ 0 \end{bmatrix}$ (spin up) and $\begin{bmatrix} 0 \\ 1 \end{bmatrix}$ (spin down). The normalized eigenvectors of x-spin are $1/\sqrt{2}\begin{bmatrix} 1 \\ 1 \end{bmatrix}$ (spin up) and $1/\sqrt{2}\begin{bmatrix} -1 \\ 1 \end{bmatrix}$ (spin down). Suppose an electron is in an eigenstate of z-spin up. Calculate the probability that an x-spin measurement will give x-spin up. Suppose an electron is in the spin state represented by $1/5\begin{bmatrix} 3 \\ 4 \end{bmatrix}$.

What is the probability that a measurement of z-spin will yield spin up? What is the probability that a measurement of x-spin will yield spin up? Suppose an electron is in the spin state represented by $1/5\begin{bmatrix} 3 \\ 4i \end{bmatrix}$. What is the probability that it will have z-spin up? x-spin up?

Time Evolution of Quantum States

We now know how the physical states of quantum systems are represented by vectors in a linear vector space and how predictions for measurements are derived from the vectors. The only question which remains is how the vectors which represent the states evolve through time. There are two ways to think about the mathematics of time evolution in quantum theory, two ways of parsing the same equations. According to one, the so-called *Schrödinger picture*, the state vectors describing the systems evolve through time. According to the other, the *Heisenberg picture*, the state vectors remain unchanged but the operators representing observables evolve. Since the Schrödinger picture is the more intuitive of the two, we will consider it exclusively.

Time evolution of state vectors is determined by another linear operator, the *Hamiltonian* of the system. The exact nature of the Hamiltonian operator need not concern us. Suffice it to say that the Hamiltonian function for a system is already defined in classical mechanics, and the quantum mechanical Hamiltonian operator can often be recovered from the classical function by replacing the position and momentum variables in the classical expression with the corresponding quantum mechanical position and momentum operators. (This can be a tricky procedure since the classical variables commute while the quantum operators do not, so one does not know whether the classical expression xp should become XP or PX or something else, where P and X are the momentum and position operators.)

Once we have the Hamiltonian operator, H, time evolution is *almost always* determined by Schrödinger's equation:

$$\frac{ih}{2\pi}\frac{d}{dt}|\psi(t)>=H|\psi(t)>,$$

where $|\psi(t)>$ represents the state of the system as a function of time and h is Planck's constant. Thus H determines the infinitesimal changes of the state vectors through time. If we need to know how the state vector has changed over some finite period of time then we need to add up a sequence of such infinitesimal changes. Formally, this is done by exponentiating the Hamiltonian operator to form an operator $U(t) = e^{-2\pi iHt/h}$. The operator $U(t)$ advances a state vector ahead in time: $|\psi(t)> = U(t)|\psi(0)>$.

We noted that H is a linear operator. Indeed, it is a Hermitian operator, and so one may naturally wonder what observable the operator is associated with. In most cases (although not all) the Hamiltonian represents the total energy of a system.

It is the linearity of H which underlies the most famous interpretational difficulties of quantum theory, and which forces orthodox quantum theory to postulate a second sort of temporal evolution. Recall that the linearity of H implies that

$$H\big(|\psi>+|\phi>\big)=H|\psi>+H|\phi>.$$

The same linearity caries over to the operator $U(t)$, so

$$U(t)\big(|\psi(0)>+|\phi(0)>\big)=U(t)|\psi(0)>+U(t)|\phi(0)>=|\psi(t)>+|\phi(t)>.$$

Schrödinger used this fact in his famous problem of the cat. Suppose, for example, we set up a device which consists of a z-spin detector coupled to a vial of poisonous gas in such a way that if a particle fed in registers spin up the vial is broken and the cat dies, while if the z-spin registers down the cat survives. Formally, this means that the time evolution operator is such that the composite system which consists of the device in its initial state plus a z-spin up particle becomes a system with a dead cat while a system made up of the device and a spin-down particle becomes a system with a live cat. If the experiment takes 10 seconds, schematically the situation is:

$$U(10)|\uparrow_z> |\text{ready}> = |\text{dead}>$$
$$U(10)|\downarrow_z> |\text{ready}> = |\text{alive}>,$$

where | $\uparrow_z$> |ready > represents the quantum state of the system composed of a z-spin up particle and the device in its ready state and | $\downarrow_z$> |ready > is the quantum state of a z-spin down particle together with the device in its ready state. What happens if we feed a particle with *x-spin* up into the device? First we note that the spin state of the particle can be expressed as a linear combination of states of definite z-spin. Recall that the eigenstate with x-spin up is $1/\sqrt{2}\begin{bmatrix}1\\1\end{bmatrix}$. This is just $1/\sqrt{2}\begin{bmatrix}1\\0\end{bmatrix}+1/\sqrt{2}\begin{bmatrix}0\\1\end{bmatrix}$. Since $\begin{bmatrix}1\\0\end{bmatrix}$ is the eigenstate of z-spin up | $\uparrow_z$> and $\begin{bmatrix}0\\1\end{bmatrix}$ is the eigenstate of z-spin down | $\downarrow_z$>, we can write the eigenstate of x-spin up as $1/\sqrt{2}|\uparrow_z>+1/\sqrt{2}|\downarrow_z>$. Using the linearity of the time evolution operator we can now calculate what the state of the cat will be 10 seconds after we feed in the x-spin up particle:

$$
\begin{aligned}
U(10)|\uparrow_x> |\text{ready}> &= U(10)(1/\sqrt{2}\ |\uparrow_z> + 1/\sqrt{2}\ |\downarrow_z>)\ |\text{ready}> \\
&= U(10)(1/\sqrt{2}\ |\uparrow_z>\ |\text{ready}> + 1/\sqrt{2}\ |\downarrow_z>\ |\text{ready}>) \\
&= U(10)1/\sqrt{2}\ |\uparrow_z>\ |\text{ready}> + U(10)1/\sqrt{2}\ |\downarrow_z>\ |\text{ready}> \\
&= 1/\sqrt{2}\ U(10)\ |\uparrow_z>\ |\text{ready}> + 1/\sqrt{2}\ U(10)|\downarrow_z>\ |\text{ready}> \\
&= 1/\sqrt{2}\ |\text{dead}> + 1/\sqrt{2}\ |\text{alive}>.
\end{aligned}
$$

That is, if we feed an x-spin up particle into this device and if time evolution is always in accord with Schrödinger's equation, then at the end of 10 seconds the wave-function describing the cat will be one according to which the cat is neither definitely alive nor definitely dead, but in a superposition of the two states.

This result was considered to be unacceptable. Cats, it is thought, simply cannot be in states in which they are neither determinately alive or determinately dead. How is one to deal with this problem?

One approach is to deny that the wave-function is a complete description of the physical state of the universe. Thus even though there may be nothing more cat-alivish than cat-deadish about the wave-function, there may be further physical facts which render the cat determinately dead or alive. This idea is exemplified in Bohm's theory.

Another approach is to simply accept that at the end of the experiment the cat is neither determinately dead or alive. The problem now is to account for our strong impression that the experiment had one particular result. Theories that pursue this approach include Everett's Many-Worlds (or Relative State) interpretation (see De Witt and Graham 1973) and the Many-Minds theory of David Albert and Barry Loewer (1988 and 1989). Orthodox quantum theory, though, adopts another tack. In the usual interpretations one denies that the cat ever gets into the state $1/\sqrt{2}$ |dead > + $1/\sqrt{2}$ | alive >. Since Schrödinger evolution leads inevitably to this state, the orthodox interpretation denies that systems always evolve in accordance with Schrödinger's equation. There is a second type of evolution which occasionally occurs and which is not governed by the Hamiltonian.

The outstanding questions for the orthodox approach are how and when the non-Schrödinger evolution takes place. The standard (albeit unsatisfactory) answer is that the second type of evolution occurs when a measurement is made on the system. When a measurement is made, the system ends up in a state which is an eigenstate of the operator for the measured quantity, with the probability for ending up in each eigenstate being the probability we have learned to calculate. Since Schrödinger evolution is completely deterministic, these unpredictable quantum jumps into eigenstates are the locus of the stochastic features of quantum theory.

One central difficulty for the orthodox interpretation is specifying under what physical conditions a measurement may be said to have occurred. If we don't know this then we don't have any account of the dynamics of the system since we don't know how its state vector will change. Suggestions for identifying when a measurement occurs include: when a macroscopic change takes place, when an irreversible change takes place, when a conscious awareness of an outcome occurs. A more recent suggestion of Ghirardi, Rimini, and Weber (1986) has it that non-Schrödinger evolution takes place entirely at random, causing from time to time the position of a particle to become localized.

The non-Schrödinger evolution is commonly called the collapse of the wave-function, since a wave-function which is initially "spread out" among the various eigenstates of an operator (i.e. is a linear combination of many such eigenstates) jumps to being one specific eigenstate.

We have already seen examples of the collapse of the wave-function. When we sent light through a polarizer, the emerging light was all polarized in the direction of the polarizer it passed. That is, the emerging light was in an eigenstate of polarization in the direction that the polarizer is set. Thus the collapse of the wave-function is also commonly used to prepare quantum systems in a known state. (In non-collapse theories such as Bohm's the wave function never collapses, and so the collapse must be explained as a purely epistemic change. This is discussed in appendix B.)

According to standard quantum theory, Schrödinger evolution and wave collapse on measurement are the only ways that the state vectors of systems change. We have therefore completed our account of standard quantum theory.

Composite Systems

Since we are concerned with the quantum state of pairs of particles, a few words about composite systems are in order. Suppose we have two systems, L and R, each with its associated Hilbert space of state vectors H_L and H_R. What is the vector space of quantum states of the composite system L + R?

The composite vector space *includes* the Cartesian product of H_L and H_R. That is, if $|V_i>_L$ is any vector in H_L and $|V_j>_R$ is any vector in H_R, the composite system can be in the product state $|V_i>_L \times |V_j>_R$, a state in which the statistical predictions for the left-hand system are given by $|V_i>_L$, the predictions for the right-hand system by $|V_j>_R$, and in which the two systems are statistically independent of one another (i.e. observations on one system would not result in altered predictions about the other). In quantum formalism the indication of the product is omitted, so the product state is written simply as $|V_i>_L |V_j>_R$, as we did above with the particle and detector states.

The essential thing to note is that the Hilbert space for the composite system is not exhausted by these product states. Any linear vector space contains *all of the linear combinations of the vectors in it*. So the vector space for the composite system contains not only, e.g., $|V_1>_L |V_1>_R$ and $|V_2>_L |V_2>_R$, but also $1/\sqrt{2} \: |V_1>_L |V_1>_R + 1/\sqrt{2} \: |V_2>_L |V_2>_R$.

The significance of vectors such as $1/\sqrt{2} \: |V_1>_L |V_1>_R + 1/\sqrt{2} \: |V_2>_L |V_2>_R$ is that they *cannot be written as the product of a vector in H_L and a vector in H_R*. The quantum state of the composite system cannot always be considered to be merely the logical sum of the individual quantum states of its components. This is the source of the holism and interconnectedness of quantum states, and hence the ultimate source of the violations of Bell's inequality.

We can now write down explicitly the entangled quantum state of our two particles (instead of photons we will use electrons). If we have a pair of electrons then the vector space of states which describes the spin state of the pair is constructed from the vector spaces for the spins of the electron individually. The spin vector space for the left-hand particle includes such vectors as $| \uparrow_z>_L, | \downarrow_z>_L, | \uparrow_x>_L, | \downarrow_x>_L$, etc. (You know how to write these states as matrices.) In fact, as we know from our discussion of spin, any vector in the left-hand vector space can be written as $\alpha | \uparrow_z>_L + \beta | \downarrow_z>_L$, since the vector space is spanned by the eigenvectors of any operator. (We could just as well have used the eigenstates of x- or y-spin). Similarly, the spin vector space of the right-hand particle contains $| \uparrow_z>_R, | \downarrow_z>_R$, and all of their linear combinations. We can form a *basis* for the vector space of the composite system by taking the products of the basis vectors of the component systems. So the spin-space of the composite system is a four-dimensional vector space which is spanned by the vectors $|\uparrow_z>_L |\uparrow_z>_R$, $|\uparrow_z>_L |\downarrow_z>_R$, $|\downarrow_z>_L |\uparrow_z>_R$, and $|\downarrow_z>_L |\downarrow_z>_R$. Any linear combination of these four vectors is a vector in the space. The spin state of pairs of electrons which violate the Bell inequality is

$$1/\sqrt{2} \, |\uparrow_z>_L |\downarrow_z>_R \; -1/\sqrt{2} \, |\downarrow_z>_L |\uparrow_z>_R$$

This is not the Cartesian product of vectors in H_L and H_R, so neither electron has any definite spin.

But what if we ask about the operator which represents the *product* of the z-spin of particle L and the z-spin of particle R? Note that $|\uparrow_z>_L |\downarrow_z>_R$ is an eigenstate of the product operator with eigenvalue –1 since if the pair is in this state the left particle has z-spin +1 and the right z-spin –1. Similarly, $|\downarrow_z>_L |\uparrow_z>_R$ is an eigenstate of the same operator, also with eigenvalue –1 (the operator is therefore degenerate). The linearity of the operator guarantees that the linear sum $1/\sqrt{2} \, |\uparrow_z>_L |\downarrow_z>_R \; -1/\sqrt{2} \, |\downarrow_z>_L |\uparrow_z>_R$ is also an eigenstate, with eigenvalue –1. Thus *the z-spins of the particles are certain to have opposite values even though neither particle has a determinate z-spin.*

The vector space for spin in an three-particle system is an eight-dimensional space, and is briefly described in Mermin's (1990b) article on the GHZ scheme.

Doing calculations on the composite system is a bit complicated since the spin operators for the left and right particles individually are degenerate. Furthermore, calculations are usually done in a different basis than the one I have written down. I will therefore assert without proof the following fact: our choice of z-spin in writing down the entangled state was arbitrary. The state is symmetric in space, so $1/\sqrt{2} \, |\uparrow_x>_L |\downarrow_x>_R - 1/\sqrt{2} \, |\downarrow_x>_L |\uparrow_x>_R$ or $1/\sqrt{2} \, |\uparrow_y>_L |\downarrow_y>_R \; -1/\sqrt{2} \, |\downarrow_y>_L |\uparrow_y>_R$ would represent the very same state.

References

Aharonov, Y. and Albert, D. 1981: Can We Make Sense of the Measurement Process in Relativistic Quantum Mechanics? *Physical Review Letters*, D 24, 359–70.

Aharonov, Y. and Albert, D. 1984: Is the Familiar Notion of Time-Evolution Adequate for Quantum-Mechanical Systems? Part II: Relativistic Considerations. *Physical Review*, D 29, 228–34.

Albert, D. 1992: *Quantum Mechanics and Experience*. Cambridge, Mass.: Harvard University Press.

Albert, D. and Loewer, B. 1988: Interpreting the Many-Worlds Interpretation. *Synthese*, 77, 195–213.

Albert, D. and Loewer, B. 1989: Two No-Collapse Interpretations of Quantum Mechanics. *Nous*, 12, 121–38.

Albert, D. and Loewer B. 1996: The Tails of Schödinger's Cat. In R. Clifton (ed.), *Perspectives on Quantum Reality*, Dordrecht: Kluwer, 81–92.

Allori, V., Goldstein, D., Tumulka, R., and Zanghì, N. 2008: On the Common Structure of Bohmian Mechanics and the Ghirardi–Rimini–Weber Theory. *British Journal for the Philosophy of Science*, 58, 353–89.

Allori, V., Goldstein, D., Tumulka, R., and Zanghì, N. 2011: Many-Worlds and Schrödinger's First Quantum Theory. *British Journal for the Philosophy of Science*, 62(1), 1–27.

Aspect, A., Dalibard, J., and Roger, G. 1982: Experimental Tests of Bell's Inequalities Using Time-varying Analyzers. *Physical Review Letters*, 49, 1804–7.

Beckmann, P. 1987: *Einstein Plus Two*. Boulder, Colorado: Golem Press.

Bell, J. S. 1987: *Speakable and Unspeakable in Quantum Mechanics*. Cambridge: Cambridge University Press.

Bell, J. S. 1990: Against "Measurement". In A. Miller (ed.), *Sixty-two Years of Uncertainty: Historical, Philosophical and Physical Inquiries into the Foundations of Quantum Mechanics*, New York: Plenum, 17–31.

Quantum Non-Locality and Relativity: Metaphysical Intimations of Modern Physics,
Third Edition. Tim Maudlin.
© 2011 Tim Maudlin. Published 2011 by Blackwell Publishing Ltd.

Bohm, D. 1952a: A Suggested Interpretation of the Quantum Theory in Terms of "Hidden Variables": Part I. *Physical Review*, 89, 166–79.

Bohm, D. 1952b: A Suggested Interpretation of the Quantum Theory in Terms of "Hidden Variables": Part II. *Physical Review*, 89, 180–93.

Bohm, D. 1957: *Causality and Chance in Modern Physics*. New York: Harper.

Bohm, D. and Hiley, B. J. 1991: On the Relativistic Invariance of a Quantum Theory Based on Beables. *Foundations of Physics*, 21, 243–50.

Bohr, N. 1935: Can Quantum Mechanical Description of Reality Be Considered Complete? *Physical Review*, 48, 696–702.

Bohr, N. 1958: *Atomic Physics and Human Knowledge*. New York: John Wiley & Sons.

Born, M. 1971: *The Born–Einstein Letters*, trans. I. Born. New York: Walker.

Butterfield, J. 1989: A Space-Time Approach to the Bell Inequality. In J. Cushing and E. McMullin, *Philosophical Consequences of Quantum Theory: Reflections on Bell's Theorem*. Notre Dame: University of Notre Dame Press, 114–44.

Cartwright, N. and Jones, M. 1991: How to Hunt Quantum Causes. *Erkenntnis*, 35, 205–31.

Clarke, C. J. S. 1977: Time in General Relativity. In J. Earman, C. Glymour and J. Stachel (eds.), *Foundations of Space-Time Theory*, Minnesota Studies in the Philosophy of Science, vol. VII, Minneapolis: University of Minnesota Press, 94–108.

Clauser, J. F. and Horne, M. A. 1974: Experimental Consequences of Objective Local Theories. *Physical Review*, D 10, 526–35.

Clauser, J. F., Horne, M. A., Shimony, A., and Holt, R. A. 1969: Proposed Experiment to Test Local Hidden-Variable Theories. *Physical Review Letters*, 26, 880–4.

Clifton, R. K., Redhead, M. L. G., and Butterfield, J. N. 1991a: Generalization of the Greenberger-Horne-Zeilinger Algebraic Proof of Nonlocality. *Foundations of Physics*, 21, 149–84.

Clifton, R. K., Redhead, M. L. G., and Butterfield, J. N. 1991b: A Second Look at a Recent Algebraic Proof of Nonlocality. *Foundations of Physics Letters*, 4, 395–403.

Conway, J. and Kochen, S. 2006: The Free Will Theorem. *Foundations of Physics*, 36, 1441–73.

Conway, J. and Kochen, S. 2009: The Strong Free Will Theorem. *Notices of the American Mathematical Society*, 56, 226–32.

Cramer, J. 1980: Generalized Absorber Theory and the Einstein–Podolsky–Rosen Paradox. *Physical Review*, D 22, 362–76.

Cramer, J. 1986: The Transactional Interpretation of Quantum Mechanics. *Review of Modern Physics*, 58, 647–87.

De Witt, B. S. and Graham, N. 1973: *The Many-Worlds Interpretation of Quantum Mechanics*. Princeton: Princeton University Press.

Dieks, D. 1985: On the Covariant Description of Wavefunction Collapse. *Physics Letters*, 108A, 379–83.

Durr, D., Goldstein, S., and Zanghì, N. 1992: Quantum Equilibrium and the Origin of Absolute Uncertainty. *Journal of Statistical Physics*, 67, 843–907.

Earman, J. 1986: *A Primer on Determinism*. Dordrecht: Reidel.

Earman, J. 1989: *World Enough and Space-Time: Absolute Versus Relational Theories of Space and Time*. Cambridge, Mass.: Bradford.

Einstein A., Podolsky, B., and Rosen, N. 1935: Can Quantum Mechanical Description of Reality Be Considered Complete? *Physical Review*, 47, 777–80.

Feinberg, G. 1967: Possibility of Faster-than-Light Particles. *Physical Review*, 159, 1089–105.

Fine, A. 1982: Some Local Models for Correlation Experiments. *Synthese*, 50, 279–94.

Fine, A. 1986: *The Shaky Game*. Chicago: University of Chicago Press.

Fine, A. 1989a: Do Correlations Need To Be Explained? In J. Cushing and E. McMullin (eds), *Philosophical Consequences of Quantum Theory: Reflections on Bell's Theorem*, Notre Dame: University of Notre Dame Press, 175–94.

Fine, A. 1989b: Correlations and Efficiency: Testing the Bell Inequalities. *Foundations of Physics*, 19, 453–78.

Fleming, G. 1985: Towards a Lorentz Invariant Quantum Theory of Measurement. (Penn State Preprint) unpublished MS.

Fleming, G. 1986: On a Lorentz Invariant Quantum Theory of Measurement. In D. Greenberger (ed.), *New Techniques and Ideas in Quantum Measurement Theory*, New York: New York Academy of Sciences, no. 480, 574–5.

Fleming, G. and Bennett, H. 1989: Hyperplane Dependence in Relativistic Quantum Mechanics. *Foundations of Physics*, 19, 231–67.

Geroch, R. 1978: *General Relativity from A to B*. Chicago: University of Chicago Press.

Ghirardi, G. C., Rimini, A., and Weber, T. 1986: Unified Dynamics for Microscopic and Macroscopic Systems. *Physical Review*, D 34, 470–91.

Giovannini, N. 1983: Relativistic Kinematics and Dynamics: a New Group Theoretical Approach. *Helvetica Physica Acta*, 56, 1002–23.

Goldstein, S. 1987: Stochastic Mechanics and Quantum Theory. *Journal of Statistical Physics*, 47, 645–67.

Goldstein, S., Tausk, D., Tumulka, R., and Zanghì, N. 2010: What Does the Free Will Theorem Actually Prove? *Notices of the American Mathematical Society*, 57(11), 1451–3.

Greenberger, D. M., Horne, M., and Zeilinger, A. 1989: Going Beyond Bell's Theorem. In M. Kafatos (ed.), *Bell's Theorem, Quantum Theory, and Conceptions of the Universe*, Dordrecht: Kluwer Academic, 69–72.

Hawking, S. W. and Ellis, G. F. R. 1973: *The Large-Scale Structure of Space-Time*. Cambridge: Cambridge University Press.

Healey, R. 1992: Chasing Quantum Causes: How Wild Is the Goose? *Philosophical Topics*, 20, 181–204.

Hellman, G. 1982: Stochastic Einstein-locality and the Bell Theorems. *Synthese*, 53, 461–504.

Hellwig, K. E. and Crause, K. 1970: Formal Description of Measurements in Local Quantum Field Theory. *Physical Review*, D 1, 566–71.

Herbert, N. 1985: *Quantum Reality*. Garden City, New York: Anchor Press Doubleday.

Howard, D. 1985: Einstein on Locality and Separability. *Studies in the History and Philosophy of Science*, 16, 171–201.

Howard, D. 1989: Holism, Separability, and the Metaphysical Implications of the Bell Experiments. In J. Cushing and E. McMullin (eds), *Philosophical Consequences of Quantum Theory: Reflections on Bell's Theorem*, Notre Dame: University of Notre Dame Press, 224–53.

Jarrett, J. 1984: On the Physical Significance of the Locality Conditions in the Bell Arguments. *Nous*, 18, 569–89.

Jones, M. 1991: Some Difficulties for Clifton, Redhead, and Butterfield's Recent Proof of Nonlocality. *Foundations of Physics Letters*, 4, 385–94.

Jones, M. and Clifton, R. 1993: Against Experimental Metaphysics. In *Midwest Studies in Philosophy*, vol. XVIII, Notre Dame: University of Notre Dame Press, 295–316.

Károlynázy, F., Frenkel, A., and Lukács, B. 1986: On the Possible Role of Gravity in the Reduction of the Wave Function. In R. Penrose and C. Isham (eds), *Quantum Concepts in Space and Time*, Oxford: Oxford University Press, 109–28.

Lewis, D. 1979: Counterfactual Dependence and Time's Arrow. *Nous*, 13, 455–76.

Lewis, D. 1986: *On the Plurality of Worlds*. Oxford: Basil Blackwell.

Lewis, P. 1995: GRW and the Tails Problem. *Topoi*, 14, 23–33.

Lo, T. K. and Shimony, A. 1981: Proposed Molecular Test of Local Hidden-Variables. *Physical Review*, 23, 3003–12.

Maudlin, T. 1990: Time Travel and Topology. In A. Fine, N. Forbes and L. Wessels (eds), *PSA 1990*, vol. 1, East Lansing: Philosophy of Science Association, 304–15.

Maudlin, T. 1992: Bell's Inequality, Information Transmission, and Prism Models. In D. Hull, M. Forbes, and K. Okruhlik (eds), *PSA 1992*, vol. 1, East Lansing: Philosophy of Science Association, 404–17.

Maudlin, T. 1995: Three Measurement Problems. *Topoi*, 14, 7–15.

Maudlin, T. 1996: Space-Time in the Quantum World. In J. Cushing, A. Fine, and S. Goldstein (eds.), *Bohmian Mechanics and Quantum Theory: An Appraisal*, Dordrecht: Kluwer, 285–307.

Maudlin, T. 1997: Descrying the World in the Wavefunction. *The Monist*, 80, 2–23.

Maudlin, T. 1998: Part and Whole in Quantum Mechanics. In E. Castellani (ed.), *Interpreting Bodies: Classical and Quantum Objects in Modern Physics*, Princeton: Princeton University Press, 46–60.

Maudlin, T. 2007: A Modest Proposal Concerning Laws, Counterfactuals, and Explanations. Chapter 1 in *The Metaphysics Within Physics*, Oxford: Oxford University Press.

Mermin, D. 1981: Quantum Mysteries for Anyone. *Journal of Philosophy*, 78, 397–408.

Mermin, D. 1989: Can You Help Your Team Tonight By Watching on TV?: More Experimental Physics From Einstein, Podolsky, and Rosen. In J. Cushing and E. McMullin (eds), *Philosophical Consequences of Quantum Theory: Reflections on Bell's Theorem*, Notre Dame: University of Notre Dame Press, 38–59.

Mermin, D. 1990a: The Philosophical Writings of Neils Bohr. In D. Mermin, *Boojums All the Way Through*, Cambridge: Cambridge University Press, 186–9.

Mermin, D. 1990b: What's Wrong With These Elements of Reality? *Physics Today*, June 1990, 9–11.

Minkowski, H. 1952: Space and Time. In H. A. Lorentz, A. Einstein, H. Minkowski, and H. Weyl, *The Principle of Relativity: A Collection of Original Memoirs on the Special and General Theory of Relativity*, Trans. W. Perrett and G. B. Jeffrey, New York: Dover, 75–91.

Neurath, O. 1959: Protocol Sentences. In A. J. Ayer (ed.), *Logical Positivism*, New York: Free Press, 199–208.

Penrose, R. 1986: Gravity and State Vector Reduction. In R. Penrose and C. Isham (eds), *Quantum Concepts in Space and Time*, Oxford: Oxford University Press, 129–46.

Perle, P. 1986: Models for Reduction. In R. Penrose and C. Isham (eds), *Quantum Concepts in Space and Time*, Oxford: Oxford University Press, 84–108.

Redhead, M. 1987: *Incompleteness, Nonlocality, and Realism: A Prolegomenon to the Philosophy of Quantum Mechanics*. Oxford: Clarendon Press.

Redhead, M. 1989: Nonfactorizability, Stochastic Causality, and Passion-at-a Distance. In J. Cushing and E. McMullin (eds), *Philosophical Consequences of Quantum Theory: Reflections on Bell's Theorem*, Notre Dame: University of Notre Dame Press, 145–53.

Reichenbach, H. 1956: *The Direction of Time*. Berkeley: University of California Press.

Schrödinger, E. 1935: The Present Situation in Quantum Mechanics. Trans. John Trimmer, in J. Wheeler and W. Zurek (eds), *Quantum Theory and Measurement*. Princeton: Princeton University Press (1981), 152–67.

Schweber, S. 1962: *An Introduction to Relativistic Quantum Field Theory*. New York: Harper Row.

Sharp, W. D. and Shanks, N. 1985: Fine's Prism Models for Quantum Correlation Statistics. *Philosophy of Science*, 52, 538–64.

Shimony, A. 1989: Search for a Worldview Which Can Accommodate Our Knowledge of Microphysics. In J. Cushing and E. McMullin (eds), *Philosophical Consequences of Quantum Theory: Reflections on Bell's Theorem*, Notre Dame: University of Notre Dame Press, 25–37.

Shimony, A. 1990: An Exposition of Bell's Theorem. In A. Miller (ed.), *Sixty-two Years of Uncertainty: Historical, Philosophical, and Physical Inquiries into the Foundations of Quantum Mechanics*, New York: Plenum, 33–43.

Sklar, L. 1974: *Space, Time, and Spacetime*. Berkeley: University of California Press.

Smith, G. J. and Weingard, R. 1987: A Relativistic Formulation of the Einstein–Podolsky–Rosen Paradox. *Foundations of Physics*, 17, 149–71.

Stapp, H. 1989: Quantum Nonlocality and the Description of Nature. In J. Cushing and E. McMullin (eds), *Philosophical Consequences of Quantum Theory: Reflections on Bell's Theorem*, Notre Dame: University of Notre Dame Press, 154–74.

Teller, P. 1989: Relativity, Relational Holism, and the Bell Inequalities. In J. Cushing and E. McMullin (eds), *Philosophical Consequences of Quantum Theory: Reflections on Bell's Theorem*, Notre Dame: University of Notre Dame Press, 208–23.

Toner, B. F. and Bacon, D. 2003: Communication Cost of Simulating Bell Correlations. *Physical Review Letters*, 91, 187904.

Tumulka, R. 2006: A Relativistic Version of the Ghirardi–Rimini–Weber Model. *Journal of Statistical Physics*, 125, 821–40.

Tumulka, R. 2007: Comment on "The Free Will Theorem." *Foundations of Physics*, 34, 186–97.

Tumulka, R. 2009: The Point Processes of the GRW Theory of Wave Function Collapse. *Reviews in Mathematical Physics*, 21, 155–227.

van Fraassen, B. 1980: *The Scientific Image*. Oxford: Oxford University Press.

van Fraassen, B. 1982: The Charybdis of Realism: Epistemological Implications of Bell's Inequality. *Synthese*, 52, 25–38; reprinted with addenda in J. Cushing and E. McMullin (eds), *Philosophical Consequences of Quantum Theory: Reflections on Bell's Theorem*, Notre Dame: University of Notre Dame Press, 97–113.

von Neumann, J. 1955: *Mathematical Foundations of Quantum Mechanics*. Princeton: Princeton University Press.

Wheeler, J. and Feynman, R. 1945: Classical Electrodynamics in Terms of Direct Interparticle Action. *Reviews of Modern Physics*, 21, 425–33.

Yurke, B. and Stoler, D. 1992a: Bell's-inequality Experiments Using Independent-particle Sources. *Physical Review*, A 46, 2229–34.

Yurke, B. and Stoler, D. 1992b: Einstein–Podolsky–Rosen Effects from Independent Particle Sources. *Physical Review Letters*, 68, 1251–4.

Index

Quantum Non-Locality and Relativity: Metaphysical Intimations of Modern Physics, Third Edition. Tim Maudlin.
© 2011 Tim Maudlin. Published 2011 by Blackwell Publishing Ltd.

Printed in the United States
By Bookmasters